32nd Annual CSP Workshop: Recent Developments in Computer Simulation Studies in Condensed Matter Physics (CSP 2019)

AA007466

Journal of Physics: Conference Series Volume 2122

Athens, Georgia, USA
18-22 February

ISBN: 978-1-7138-5093-9
ISSN: 1742-6588

Printed from e-media with permission by:

Curran Associates, Inc.
57 Morehouse Lane
Red Hook, NY 12571

Some format issues inherent in the e-media version may also appear in this print version.

Institute of Physics
Dirac House, Temple Back
Bristol BS1 6BE UK

Phone: 44 1 17 929 7481
Fax: 44 1 17 920 0979

techtracking@iop.org

Additional copies of this publication are available from:

Curran Associates, Inc.
57 Morehouse Lane
Red Hook, NY 12571 USA
Phone: 845-758-0400
Fax: 845-758-2633
Email: curran@proceedings.com
Web: www.proceedings.com

TABLE OF CONTENTS

Preface

Thirty three years ago, because of the dramatic increase in the power and utility of computer simulations, The University of Georgia formed the first institutional unit devoted to the application of simulations in research and teaching: The Center for Simulational Physics. Then, as the international simulations community expanded further, we sensed the need for a meeting place for both experienced simulators and newcomers to discuss inventive algorithms and recent results in an environment that promoted lively discussion. As a consequence, the Center for Simulational Physics established an annual workshop series on *Recent Developments in Computer Simulation Studies in Condensed Matter Physics.* This year's highly interactive workshop was the 32nd in the series marking our efforts to promote high quality research in simulational physics. The continued interest shown by the scientific community amply demonstrates the useful purpose that these meetings have served.

The latest workshop was held at The University of Georgia from February 18-22, 2019. These Proceedings provide a "status report" on a number of important topics. This on-line "volume" is published with the goal of timely dissemination of the material to a wider audience.

These Proceedings contain both invited papers and contributed presentations on problems in both classical and quantum condensed matter physics. The Workshop was prefaced by a special tutorial presented by colleagues from Oak Ridgr National Laboratory on a powerful software suite: OWL (Oak Ridge Wang-Landau). The first manuscript in this Proceedings is devoted to this tutorial material. The Workshop topics, as usual, ranged from hard and soft condensed matter to biologically inspired problems and purely methodological advances. We hope that readers will benefit from specialized results as well as profit from exposure to new algorithms, methods of analysis, and conceptual developments.

D. P. Landau
M. Bachmann
S. P. Lewis
H.-B. Schüttler

Peer review declaration

All papers published in this volume of Journal of Physics: Conference Series have been peer reviewed through processes administered by the Editors. Reviews were conducted by expert referees to the professional and scientific standards expected of a proceedings journal published by IOP Publishing.

- **Type of peer review:** *Single blind*
- **Conference submission management system:** *Manuscripts were submitted electronically to the Workshop organizers*
- **Number of submissions received:** *10*
- **Number of submissions sent for review:** *10*
- **Number of submissions accepted:** *10*
- **Acceptance Rate (Number of Submissions Accepted / Number of Submissions Received X 100):** *100*
- **Average number of reviews per paper:** *1*
- **Total number of reviewers involved:** *1*
- **Any additional info on review process:** *The manuscripts were all of high- to very high quality*
- **Contact person for queries:** *David P. Landau, Center for Simulational Physics*

Hands-on with OWL: the Oak–Ridge Wang–Landau Monte Carlo software suite

Ying Wai Li[1, 2], Krishna Chaitanya Pitike[3], Markus Eisenbach[1] and Valentino R. Cooper[3] *

[1] National Center for Computational Sciences, Oak Ridge National Laboratory, Oak Ridge, TN 37831, U.S.

[2] Computer, Computational, and Statistical Sciences Division, Los Alamos National Laboratory, Los Alamos, NM 87544, U.S.

[3] Condensed Matter Theory Group, Oak Ridge National Laboratory, Oak Ridge, TN 37831, U.S.

E-mail: `yingwaili@lanl.gov`

Abstract.
The Oak–Ridge Wang–Landau (OWL) package is an open-source scientific software specialized for large-scale, Monte Carlo simulations for the study of materials properties at finite temperature. In this paper, we discuss the main features and capabilities of OWL, followed by detailed descriptions of building and running the code. The readers will be guided through the usage and functionality of the code with a few hands-on examples. This paper is based on a tutorial on OWL given at the 32nd Center for Simulational Physics Workshop on Recent Developments in Computer Simulation Studies in Condensed Matter Physics.

1. Open source repository

OWL is open-sourced under the BSD 3-clause license. The latest version of the code can be found at: `https://github.com/owl-suite/OWL`. The discussion in this paper refers to Version `v0.0-alpha` of the code released on February 18, 2019.

2. What is OWL?

OWL is an acronym for "**O**ak–Ridge **W**ang–**L**andau" (or "**O**pen–source **W**ang–**L**andau"). It is a scientific software to perform large-scale, first-principles based statistical mechanics simulations for the study of finite temperature materials properties. Originally developed at the Oak Ridge National Laboratory, OWL was first intended for the studies of real-world materials and complex systems of which the physics cannot be represented by simple model Hamiltonians. It is also intended for performing simulations that are more readily and directly comparable

*This manuscript has been authored by UT-Battelle, LLC under Contract No. DE-AC05-00OR22725 with the U.S. Department of Energy. The United States Government retains and the publisher, by accepting the article for publication, acknowledges that the United States Government retains a non-exclusive, paid-up, irrevocable, world-wide license to publish or reproduce the published form of this manuscript, or allow others to do so, for United States Government purposes. The Department of Energy will provide public access to these results of federally sponsored research in accordance with the DOE Public Access Plan (http://energy.gov/downloads/doe-public-access-plan).

with experimental observations. For example, to account for the electronic degrees of freedom and correlations in materials more accurately, first principles methods such as density functional theory (DFT) are employed for the calculation of total energies and physical properties. Monte Carlo methods, on the other hand, are used to obtain thermodynamic and finite temperature properties, which involve the sampling of a large number of configurations to obtain reasonable statistics and survey of phase space. The combination of these two components, *ab initio* and Monte Carlo methods, requires tremendous amount of computing resources due to the multiplicative effect - the computation time needed for a single *ab initio* computation multiplied by the number of Monte Carlo samples, which can easily be on the order of 10^4 and beyond. From a technical point of view, this simulation paradigm introduces multiple levels of computational parallelism that is a perfect fit for modern computer architectures with massive, heterogeneous, parallel computing units.

An important aspect of a software that will be widely applicable is its modularity. Different Monte-Carlo methods have behaviors that can be better suited for exploring different aspects of the statistical mechanics of physical systems and a well described interface will allow for the exploration of new or improved algorithms as they become available. Furthermore, our goal of facilitating the computational statistical mechanics of complex Hamiltonians, including *ab initio* calculations and recent developments in machine learning derived models, requires the flexible compatibility of the approach taken in OWL which affords the exploration of new physical systems on a wide range of computer architectures ranging from single workstations to the largest high performance computer (HPC) systems available.

As the name implies, Wang–Landau sampling [1] was the central component of OWL and was the original motivation for the development of the code. As time and research needs evolved, OWL was extended to include a collection of commonly used classical, modern and parallel Monte Carlo (MC) algorithms.

2.1. Two modes of simulations: standalone mode v.s. driver mode

OWL provides two modes of simulations: the standalone mode and the driver mode.

- The standalone mode contains the basic functionalities that encapsulate all the Monte Carlo algorithms implemented in the code, but without any interface to external codes. It is run as an independent application for simulating user-implemented model Hamiltonians.
- The driver mode interfaces with an external package. For every Monte Carlo step, OWL sends the trial configurations to the external code to calculate the total energy and other physical properties, which are then passed back to OWL to determine the acceptance of the trial move.

To date, OWL interfaces with two open-source DFT codes, Quantum Espresso (QE) [2, 3] and Locally Self-Consistent Multiple Scattering (LSMS) [4–6]. In principle, there is no restriction on the methods that the external library employs for the energy calculations. Packages other than DFT codes, such as molecular dynamics codes, can also be used. The interfaces with FERAM [7, 8] and LAMMPS [9, 10], are two such examples that are under development.

2.2. Programming model and software design

OWL is written in C++ with an object-oriented, modular software architecture. This brings several advantages in the software development. It disentangles the Monte Carlo algorithms from the physical systems to be studied. Every algorithm only needs to be implemented once-and-for-all; the same applies to the definition of the physical model Hamiltonians. This allows for flexible combinations of algorithms and physical models. As the modular design enables code reuse and avoids duplicate implementations, it also hides the implementation details from the algorithm logic, making the code cleaner and easier to extend. For example, developers do not

Figure 1. The logo for the OWL code.

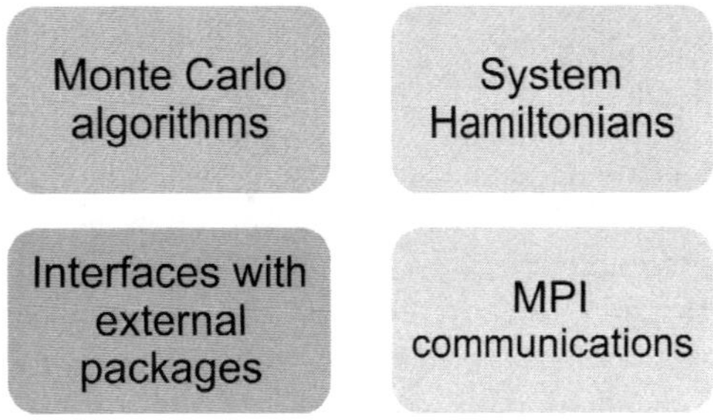

Figure 2. OWL has an object-oriented, modular design.

need to know the details of the interfaces to external codes in order to implement a new MC algorithm. Adding new features and functionality thus requires minimal effort.

OWL adopts the "MPI+X" programming model for parallelization: Message Passing Interface (MPI) [11] is a communication protocol for parallel programming. MPI provides a powerful way to easily express parallel programs through libraries. Through MPI parallelization, OWL can run efficiently on multiple computing nodes with distributed memory. We use MPI to take care of parallelization at the random walker level - each MPI rank corresponds to one random walker. By default, traditional serial code will be executed using one MPI rank assuming one random walker. The "X" component refers to additional levels of parallelism such as C++ threads or OpenMP (Open Multi-Processing) for shared memory multiprocessing [12], CUDA [13] or the like for Graphics Processing Unit (GPU) acceleration. With the MPI+X hybrid parallel programming model, MPI is used to parallelize across nodes and C++threads/OpenMP can be used to parallelize within the (multi-core) node.

In OWL, we reserve the "X" level of parallelization for physical model related computations. The external packages that OWL interfaces with usually take advantage of the "MPI+X" programming model too to perform calculations for one configuration. In this case, OWL would perform the MC simulations with one or several random walkers (depending on the algorithm), and each random walker would span several MPI ranks. OWL would serve as a computing resource manager and distribute the MPI ranks along with the "X" parallelization level to the external code for execution.

OWL uses Git for version control and code development, and the code is hosted as a GitHub repository. The "master branch" contains the most updated, development version of the code. Other branches are "feature branches" where new features and capabilities are being actively developed and are often work-in-progress. When a new implementation is completed in a feature branch, it will be tested and then merged into the master branch through a "pull request" process. This rigorous software engineering practice reduces the possibility of introducing programming bugs or incompatibilities into the master branch. Similar to other open-source code projects, we encourage users to report problems and request for features by submitting a "new issue" through GitHub.

3. Code usage

In this section, we will demonstrate the use of OWL through a few selected examples. Readers will learn about package building, how to perform serial and parallel Monte Carlo simulations for simple spin systems, as well as first-principles based Monte Carlo simulations using OWL. Updated documentation can be found on the Wiki page of the code repository: `https://github.com/owl-suite/OWL/wiki`.

3.1. Prerequisites
OWL is very light-weight and is designed with minimal dependence on external libraries and maximal portability in mind. A C++ compiler with MPI support are the only requirements to compile the standalone version of the code. OWL can be readily compiled on a variety of computing systems ranging from laptops, workstations or computer clusters, to leadership-class supercomputers.

To compile the driver mode of OWL where an external code will be used to calculate the energy and other physical quantities, the external code will need to be downloaded, compiled, and linked to OWL. We will describe the procedure in Section 3.3.2.

3.2. Downloading the code

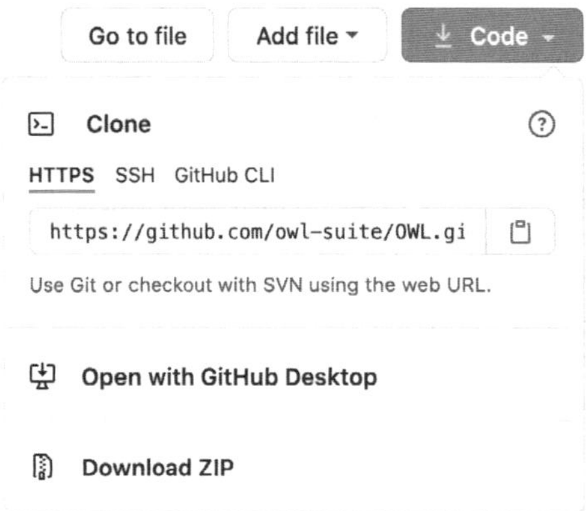

Figure 3. Downloading OWL from its GitHub site: click the green button "Code" on the OWL repository website and explore the different options. Common ways include downloading the code as a zip file (not recommended), or cloning its repository using "git" (recommended).

OWL's repository is hosted on GitHub at `https://github.com/owl-suite/OWL`. While it is possible to download the source code "as is" as a zip file (click the "Download ZIP" option under the green "Code" button on the repository site (Figure 3)), we recommend using Git to clone the repository onto your own computer by typing the following in a terminal:

```
git clone https://github.com/owl-suite/OWL.git
```

For code developers, Git provides a version control system that systematically keeps track of all the development history and versions of a software. For users, the benefit of using Git is that one may obtain the latest updates, new features, and bug fixes of the code readily using the command:

```
git pull
```

3.3. Compiling the code and linking to external codes
3.3.1. Standalone mode In the standalone mode, OWL performs the simulations using built-in simple models defined and implemented in the code. Before compiling, ensure that you have a working copy of an MPI-enabled C++ compiler (the MPI-wrapped compilers are usually named `mpicxx` or `mpic++`, with exceptions on some HPC systems).

When you first download OWL from the GitHub repository, you will see a symbolic link `architecture.inc` pointing to a string `path_to_architecture_file`. It is to remind the users to specify an appropriate architecture file for their computers. Some common computer architectures are provided in the `architecture` directory. Inside an architecture file, the MPI-enabled C++ compiler, as well as the compiler flags and link flags that would be used to compile the code, are specified. For special computer architectures or customized compiling options, the users should create their own architecture file.

Before linking to the desired architecture file, remove any previous `architecture.inc` file:

```
rm architecture.inc
```

Then recreate the symbolic link `architecture.inc` to point to the desired architecture file. For example, on a Linux machine, you may use the `generic_linux` architecture file:

```
ln -s architecture/generic_linux architecture.inc
```

For a Mac computer, the `mac_osx` architecture file is provided:

```
ln -s architecture/mac_osx architecture.inc
```

Finally, to compile the standalone version of OWL:

```
make owl
```

If the compilation finishes successfully, the executable `owl` should be created inside the `bin` directory.

3.3.2. Driver mode A powerful feature of OWL is the capability to drive an external atomistic or electronic structure code as a library for energy and physical properties calculations. This is advantageous when the system of interest cannot be defined as a simple model, especially when electronic degrees of freedom enter the Hamiltonian. In this case, OWL controls the Monte Carlo algorithms and related parameters the same way as in the standalone mode. The systems of interest can be specified using the external package's mechanism, the same way as when the calculation is carried out by that package alone.

To build OWL to drive an external code, the latter will need to be compiled first, and its source code location must be linked to OWL. In the OWL repository, we provide an example bash script, `build_QE_5.4.sh`, to illustrate how to download, compile and link the Quantum Espresso package (Version 5.4.0) to OWL as an example. This script is located in the `external_codes` directory. Assuming that working copies of the MPI-enabled C and Fortran compilers (`mpicc`, `mpif77` and `mpif90`) required to build QE are properly installed and configured, the following commands will build QE inside a directory called `q-e-qe-5.4` under `external_codes`:

```
cd external_codes
./build_QE_5.4.sh
```

It will also create a symbolic link `quantum_espresso` in OWL's top directory pointing to the top directory of QE, which is `q-e-qe-5.4` in this case.

Once QE is built, we can now build OWL in its driver mode by executing the following command from OWL's top directory:

```
make owl-qe
```

Again, if the compilation is successful, the resulting executable, `owl-qe` this time, should be created inside the `bin` directory.

3.4. Running example simulations

In the OWL repository, we provide a few example input files in the `examples` directory to illustrate the usage of different modes of the code. To run a serial Wang-Landau sampling of a 2D Ising model, we would use the standalone version of OWL using one processor (i.e., one MPI rank). From OWL's top directory, execute:

```
cd examples
cd Ising2D
mpirun -np 1 ../../bin/owl owl.input
```

More information and details about OWL's input file options are documented on the Wiki page of OWL's GitHub repository.

When running OWL in driver mode, users will have to specify additional input parameters in OWL's input file (`owl.input`) using the `PhysicalSystem` tag to indicate which external code to use. The external codes usually have their own input files to specify the physical systems, which need to be provided along with OWL's input file when OWL's driver mode is used. External packages will also have their own command line options to control a simulation/calculation; these can be specified in OWL's input file using the `PhysicalSystemCommandLine` tag.

In the `examples/CuZn` directory, we provide the input files to perform a Wang-Landau sampling for a binary alloy Copper-Zinc (CuZn), using QE to calculate the total energy using DFT. In addition to the OWL input file `owl.input`, three additional input files, `qe.input`, `Cu_pbesol_v1.2.uspp.f.upf` and `Zn_pbesol_v1.uspp.f.upf`, are required to run QE. The three QE input files should be prepared in exactly the same way as one would do for calculating this system using QE only: `qe.input` is for controlling the DFT calculations and specifying the system setup; while the `.upf` files define the pseudopotentials for the Cu and Zn atoms in the Unified Pseudopotential Format. To run the simulation, we would use the driver version of OWL, and multiple MPI ranks for QE to perform parallel DFT calculations:

```
cd examples/CuZn
mpirun -np 8 ../../bin/owl-qe owl.input
```

Interested readers are directed to consult Quantum Espresso's documentation (`https://www.quantum-espresso.org/resources/users-manual`) on the preparation of QE input files.

4. Summary and outlook

In summary, the Oak-Ridge Wang-Landau (OWL) code represents a scalable, efficient, modular approach to stochastic sampling. The goal of this project is to build a platform for a community (open source) code for exploring the properties of a material that depend on statistical averaging. The aim is that such a vehicle may not only afford more advanced materials simulations but may facilitate algorithm advancements and dissemination.

OWL contains a number of standard Monte Carlo techniques including the Wang-Landau method incorporated in a seamless fashion with the capability for easily swapping a range of potentials/Hamiltonians for the exploration of materials systems. Fundamental to the development of this code is the transferability and scalability of the code over multiple high performance computing architectures.

In essence, OWL offers a promising solution for the examination of temperature-dependent properties defined by ensemble averages using atomistic or electronic structure codes, a feat which has largely been applied to lattice models with discrete basis sets, e.g. Ising or Potts models. Initial efforts include interfaces to standard MD and electronic structure codes such as LAMMPS[9, 10], LSMS[4, 5] and Quantum Espresso[2, 3], with easily adaptable modules that could be applied to other electronic structure codes. Such developments may afford the ability to use atomistic stochastic sampling to explore temperature-dependent phase transitions as well as reactive surfaces and processes. We anticipate that further community development and application will open the door to a new era of computational research that affords direct simulation of materials properties and behavior under realistic conditions.

The features and capabilities of OWL grow steadily over time. We encourage interested users to try it out and we welcome comments, suggestions and contributions from the scientific community.

Acknowledgments

This work was sponsored by the Laboratory Directed Research and Development Program of Oak Ridge National Laboratory, managed by UT-Battelle, LLC, for the U.S. Department of Energy, together with the Oak Ridge Leadership Computing Facility, which is supported by the Office of Science of the U.S. Department of Energy under contract No. DE-AC05-00OR22725.

References

[1] Wang F and Landau D P 2001 *Phys. Rev. Lett.* **86** 2050
[2] Giannozzi P, Baroni S, Bonini N, Calandra M, Car R, Cavazzoni C, Ceresoli D, Chiarotti G L, Cococcioni M, Dabo I, Dal Corso A, de Gironcoli S, Fabris S, Fratesi G, Gebauer R, Gerstmann U, Gougoussis C, Kokalj A, Lazzeri M, Martin-Samos L, Marzari N, Mauri F, Mazzarello R, Paolini S, Pasquarello A, Paulatto L, Sbraccia C, Scandolo S, Sclauzero G, Seitsonen A P, Smogunov A, Umari P and Wentzcovitch R M 2009 *Journal of Physics: Condensed Matter* **21** 395502 (19pp) URL `http://www.quantum-espresso.org`
[3] Giannozzi P, Andreussi O, Brumme T, Bunau O, Nardelli M B, Calandra M, Car R, Cavazzoni C, Ceresoli D, Cococcioni M, Colonna N, Carnimeo I, Corso A D, de Gironcoli S, Delugas P, Jr R A D, Ferretti A, Floris A, Fratesi G, Fugallo G, Gebauer R, Gerstmann U, Giustino F, Gorni T, Jia J, Kawamura M, Ko H Y, Kokalj A, Kkbenli E, Lazzeri M, Marsili M, Marzari N, Mauri F, Nguyen N L, Nguyen H V, de-la Roza A O, Paulatto L, Ponc S, Rocca D, Sabatini R, Santra B, Schlipf M, Seitsonen A P, Smogunov A, Timrov I, Thonhauser T, Umari P, Vast N, Wu X and Baroni S 2017 *Journal of Physics: Condensed Matter* **29** 465901 URL `http://stacks.iop.org/0953-8984/29/i=46/a=465901`
[4] Wang Y, Stocks G M, Shelton W A, Nicholson D M C, Szotek Z and Temmerman W M 1995 *Phys. Rev. Lett.* **75**(15) 2867–2870 URL `https://link.aps.org/doi/10.1103/PhysRevLett.75.2867`
[5] Eisenbach M, Li Y W, Liu X, Odbadrakh O K, Pei Z, Stocks G M and Yin J 2017 Locally self-consistent multiple scattering (LSMS) URL `https://www.osti.gov//servlets/purl/1420087;https://github.com/mstsuite/lsms`
[6] Eisenbach M, Larkin J, Lutjens J, Rennich S and Rogers J H 2017 *Computer Physics Communications* **211** 2–7
[7] Nishimatsu T, Waghmare U V, Kawazoe Y and Vanderbilt D 2008 *Phys. Rev. B* **78**(10) 104104 URL `https://link.aps.org/doi/10.1103/PhysRevB.78.104104`
[8] Nishimatsu T Feram: Md simulator for bulk and thin-film ferroelectrics URL `http://loto.sourceforge.net/feram/`
[9] Plimpton S 1995 *Journal of Computational Physics* **117** 1 – 19 ISSN 0021-9991 URL `http://www.sciencedirect.com/science/article/pii/S002199918571039X`
[10] Plimpton S LAMMPS URL `http://lammps.sandia.gov`
[11] MPI official website URL `https://www.mpi-forum.org/`
[12] OpenMP official website URL `https://www.openmp.org/`
[13] CUDA official website URL `https://developer.nvidia.com/CUDA-zone`

Journal of Physics: Conference Series **2122** (2021) 012002 doi:10.1088/1742-6596/2122/1/012002

Role of electron correlations in some Weyl systems

Niraj Aryal[(1)] **and Efstratios Manousakis**[(1,2)]

[(1)]Department of Physics and National High Magnetic Field Laboratory, Florida State University, Tallahassee, FL 32306-4350, USA

[(2)]Department of Physics, University of Athens, Panepistimioupolis, Zografos, 157 84 Athens, Greece

E-mail: `manousakis@magnet.fsu.edu`

Abstract. We are discussing a model to understand previously obtained results on Weyl semimetals as realized in $MoTe_2$ using DFT and DFT+U calculations. The model is motivated from general principles and we use it to investigate the effects of Coulomb correlations originating from the localized nature of the Mo-d orbitals. We find that such correlations can eliminate or create pairs of Weyl points as the strength of the Coulomb interaction is varied. The effect of the spin-orbit coupling (SOC) is to split each Weyl point, which is assumed present in the absence of SOC, into pairs of spin-chiral partners.

1. Introduction

Among several fascinating properties of transition metal dichalcogenides (TMDs)[1, 2, 3] is that some are Weyl semimetals[4, 5, 6]. In particular the band structure of $MoTe_2$ has been predicted by density functional theory calculations to contain pairs of Weyl points (WPs) not too far away from the Fermi level. While subsequent angular resolved photoemission spectroscopic (ARPES) studies seem to confirm that prediction [7, 8, 9, 10], maps of the Fermi surface by detailed quantum oscillation experiments (QOE) seem to contradict such a scenario [11]. Subsequent detailed calculations, within the GGA+U approximation, indicate that the effect of Coulomb electron corrections is significant in order to explain both the light-polarization dependence of ARPES and the angular dependence of the frequencies observed by the QOE [12]. In addition, these calculations suggest that while the Weyl points nearest to the Fermi energy survive the inclusion of these correlations, interesting Lifshitz transitions occur which change the topology of the Fermi surface.

In the present paper, we present a rather simple analytical model based on the nature of the Weyl points in $MoTe_2$, to help understand the role of electron correlations in this material and the results obtained in Ref. [12].

First we begin by deriving the model based on the results of the GGA+U calculation presented in Fig. 1. The size of the dots in Fig. 1 gives the relative contribution of the p and d orbitals to the bands, while other orbitals have very small contribution. Notice that the two bands forming the Weyl-point far away from the Weyl-point are mostly of one of the two characters and near the Weyl-point there is hybridization. Therefore, it makes sense to construct a model containing just a pair of orbitals A and B representing the orbitals d and p respectively. The most general

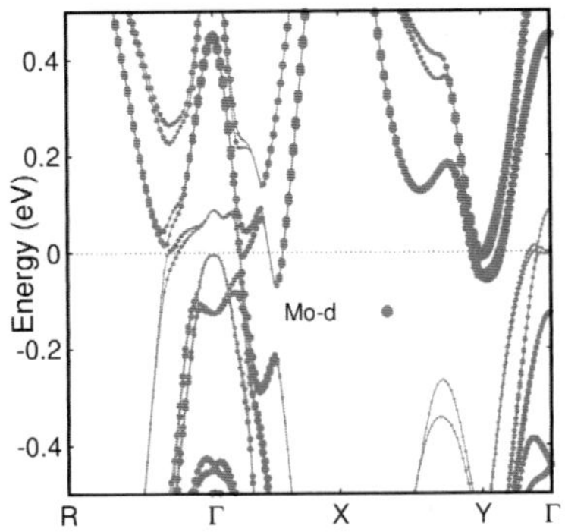

(a)

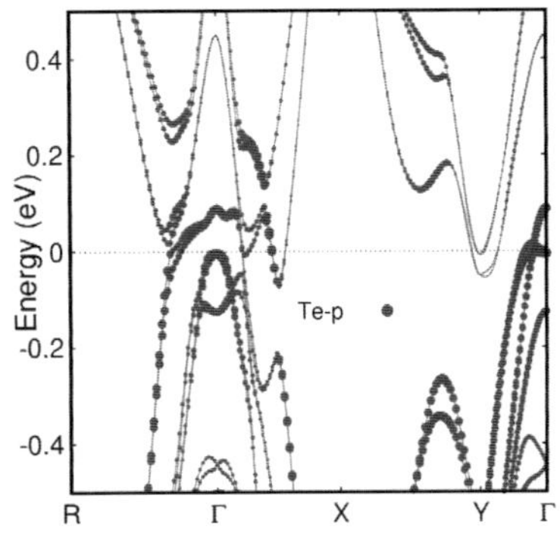

(b)

Figure 1. Band structure of γ-$MoTe_2$ showing the projection onto (a) Mo-d orbitals and (b) Te-p orbitals. The size of the dots is proportional to the percentage of the corresponding orbitals.

form of the tight-binding Hamiltonian in k-space is

$$\mathcal{H}(\mathbf{k}) = \begin{bmatrix} e_A(\mathbf{k}) & f_1(\mathbf{k}) + if_2(\mathbf{k}) \\ f_1(\mathbf{k}) - if_2(\mathbf{k}) & e_B(\mathbf{k}) \end{bmatrix}, \quad (1)$$

which can also be re-written as

$$\begin{aligned} \mathcal{H}(\mathbf{k}) &= \epsilon(\mathbf{k})\hat{I} + \vec{h}(\mathbf{k}) \cdot \vec{\tau}, \\ \epsilon(\mathbf{k}) &= (e_A(\mathbf{k}) + e_B(\mathbf{k}))/2 \\ \vec{h}(\mathbf{k}) &= (h_1(\mathbf{k}), h_2(\mathbf{k}), h_3(\mathbf{k})), \\ h_1(\mathbf{k}) &= (e_A(\mathbf{k}) - e_B(\mathbf{k}))/2, \quad h_2(\mathbf{k}) = f_1(\mathbf{k}), \quad h_3(\mathbf{k}) = f_2(\mathbf{k}), \end{aligned} \quad (2)$$

where $\vec{\tau} = (\tau_x.\tau_y, \tau_z)$ and $\tau_{x,y,z}$ are the Pauli matrices. A Weyl point can exist when there is a $\mathbf{k}_0$ which is the solution to the system of equations:

$$h_1(\mathbf{k}_0) = 0, \quad h_2(\mathbf{k}_0) = 0, \quad h_3(\mathbf{k}_0) = 0. \quad (3)$$

Each one of the equations above defines a "surface" in a d-dimensional k-space and, in principle, in $d \geq 3$ where there are d unknowns, i.e., the d-components of $\mathbf{k}_0$, there could be solutions independently of the nature of the symmetries of the Hamiltonian.

Let us ignore the SOC and add the on-site Coulomb Hubbard U term on the A orbital only:

$$\hat{H} = \sum_{\vec{k},\alpha\beta\sigma} \mathcal{H}_{\alpha\beta}(\vec{k}) C^{\dagger}_{\beta\sigma}(\vec{k}) C_{\alpha\sigma}(\vec{k}) + U \sum_{\vec{R}} \hat{n}^{(A)}_{\uparrow}(\vec{R}) \hat{n}^{(A)}_{\downarrow}(\vec{R}) \quad (4)$$

where $n^{(A)}_{\sigma}(\vec{R}) = C^{\dagger}_{A\sigma}(\vec{R}) C_{A\sigma}(\vec{R})$. In mean field factorization, we write

$$\hat{n}^{(A)}_{\uparrow}(\vec{R})\hat{n}^{(A)}_{\downarrow}(\vec{R}) \rightarrow \langle \hat{n}^{(A)}_{\uparrow}(\vec{R})\rangle \hat{n}^{(A)}_{\downarrow}(\vec{R}) + \hat{n}^{(A)}_{\uparrow}(\vec{R})\langle \hat{n}^{(A)}_{\downarrow}(\vec{R})\rangle - \langle \hat{n}^{(A)}_{\uparrow}(\vec{R})\rangle\langle \hat{n}^{(A)}_{\downarrow}(\vec{R})\rangle \quad (5)$$

and if the system is not magnetically ordered we take $\langle \hat{n}_{\uparrow}^{(A)}(\vec{R})\rangle = \langle \hat{n}_{\downarrow}^{(A)}(\vec{R})\rangle = n_A$. In this case the Hamiltonian simply becomes

$$\hat{H} = \sum_{\vec{k},\alpha\beta\sigma} \mathcal{H}'_{\alpha\beta}(\vec{k}) C^{\dagger}_{\beta\sigma}(\vec{k}) C_{\alpha\sigma}(\vec{k}) + E_0 \tag{6}$$

where E_0 is a constant and

$$\mathcal{H}'(\vec{k}) = \begin{bmatrix} e_A(\mathbf{k}) + n_A U & f_1(\mathbf{k}) + i f_2(\mathbf{k}) \\ f_1(\mathbf{k}) - i f_2(\mathbf{k}) & e_B(\mathbf{k}) \end{bmatrix} \tag{7}$$

for both up and down spins. However, Eqn. 7 is similar to Eq. 3 with the only difference is that $h_1(\mathbf{k})$ is replaced by:

$$h_1(\mathbf{k}) = \frac{e_A(\mathbf{k}) + n_A U - e_B(\mathbf{k})}{2}. \tag{8}$$

Let us discuss, the role of U, assuming that there were WP solutions before the introduction of the Hubbard term. Since, we have assumed that there is a WP solution before we introduce an non-zero U, it implies that the two surfaces in k-space defined by the equations $h_2(\mathbf{k}) = 0$ and $h_3(\mathbf{k}) = 0$ remain the same as in the case where U=0. Therefore, these two equations should define a intersection, which is a line l, which is unaffected by the addition of the Hubbard term, within mean-field treatment. For a small value of U the modified equation $h_1(\mathbf{k}) = 0$ defines a somewhat different surface which for a sufficiently small value of U should continue to intersect the line l at least in two points, the WPs. To highlight this point, that the WPs continue to exist for small value of U, although they change their relative positions, we illustrate the positions of the source and of the sink of the WPs for zero and non-zero U value in Fig. 2.

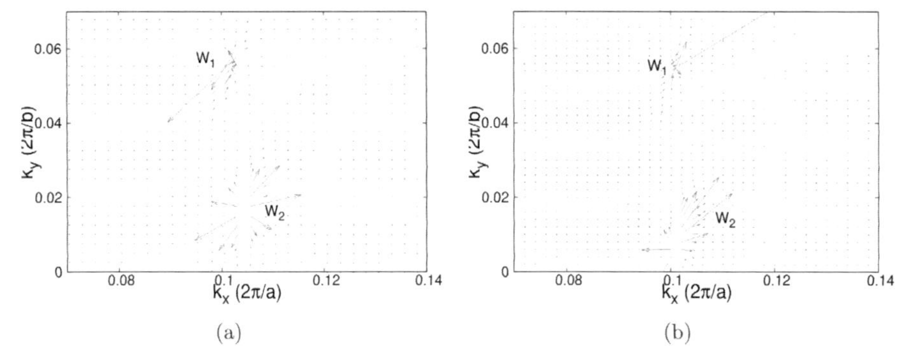

Figure 2. Vector plot of the Berry curvature on $k_z = 0$ plane showing the source and sink of the WPs for U = 0 and U = 0.75 eV respectively as obtained from DFT calculations. The relative size of the length of the vector is proportional to the magnitude of the Berry curvature. Notice the shift of the WPs in (b) as compared to (a).

We can linearize the equations $h_1(\mathbf{k}) = 0$, $h_2(\mathbf{k}) = 0$, and $h_3(\mathbf{k}) = 0$, around each of the band touching points $\mathbf{k_0}$ where all three components of $\mathbf{h}(\mathbf{k}_0)$ vanish as mentioned in Eqn. 3. A Taylor expansion around $\mathbf{k} = \mathbf{k}_0$ gives:

$$h_i(\mathbf{k}) = h_i(\mathbf{k}_0) + \sum_j v_{ij}(\mathbf{k}_0)(\mathbf{k} - \mathbf{k}_0)_j + O(\mathbf{k} - \mathbf{k}_0)^2 \tag{9}$$

where, $v_{ij} = \frac{\partial h_i}{\partial k_j}$ is the Fermi velocity tensor. Hence, the Hamiltonian near the band touching point takes the form of:

$$\mathcal{H}(\mathbf{k}) \quad = \quad \begin{bmatrix} \widetilde{h}_3(\mathbf{k}) & \widetilde{h}_1(\mathbf{k}) + i\widetilde{h}_2(\mathbf{k}) \\ \widetilde{h}_1(\mathbf{k}) - i\widetilde{h}_2(\mathbf{k}) & -\widetilde{h}_3(\mathbf{k}) \end{bmatrix} \tag{10}$$

where, $\widetilde{h}_i(\mathbf{k}) = v_{ix}k'_x + v_{iy}k'_y + v_{iz}k'_z$ and $k'_i = (\mathbf{k} - \mathbf{k_0})_i$. Equivalently, Eqn. 10 can be written as:

$$\mathcal{H}(\mathbf{k}) = \sum_{ij} \tau_i v_{ij} k'_j. \tag{11}$$

A unitary transformation that diagonalizes the velocity tensor $\mathbf{v}$ transforms Eqn. 10 into the familiar Weyl equation of the form $\vec{\tau}.\mathbf{k}$.

Therefore, we have shown that the most general form of the 2×2 Hamiltonian reduces to the Weyl Hamiltonian in the vicinity of the band touching point, even with the addition of the Hubbard term U for some range of U.

The role of spin-orbit coupling was studied in Ref. [12] and it was found that it doubles the number of Weyl pairs and from a single pseudo-spin chiral pair it produces two real spin-chiral pairs of Weyl points.

2. Acknowledgments

This work was supported in part by the U.S. National High Magnetic Field Laboratory, which is funded by NSF DMR-1157490 and the State of Florida.

References

[1] Wang Q H, Kalantar-Zadeh K, Kis A, Coleman J N and Strano M S 2012 *Nat. Nanotechnol.* **7** 699–
[2] Wilson J and Yoffe A 1969 *Adv. Phys.* **18** 193–335
[3] Mattheiss L F 1973 *Phys. Rev. B* **8**(8) 3719–3740
[4] Wang Z, Gresch D, Soluyanov A A, Xie W, Kushwaha S, Dai X, Troyer M, Cava R J and Bernevig B A 2016 *Phys. Rev. Lett.* **117**(5) 056805
[5] Soluyanov A A, Gresch D, Wang Z, Wu Q, Troyer M, Dai X and Bernevig B A 2015 *Nature* **527** 495–
[6] Chang T R, Xu S Y, Chang G, Lee C C, Huang S M *et al.* 2016 *Nat. Commun.* **7** 10639–
[7] Huang L, McCormick T M, Ochi M, Zhao Z, Suzuki M T, Arita R, Wu *et al.* 2016 *Nat. Mater.* **15** 1155–
[8] Deng K, Wan G, Deng P, Zhang K, Ding S, Wang E, Yan M, Huang *et al.* 2016 *Nat. Phys.* **12** 1105–
[9] Jiang J, Liu Z, Sun Y, Yang H, Rajamathi C, Qi Y, Yang L, Chen C, Peng H *et al.* 2017 *Nat. Commun.* **8** 13973–
[10] Tamai A, Wu Q S, Cucchi I, Bruno F Y, Riccò S, Kim T K, Hoesch M, Barreteau C *et al.* 2016 *Phys. Rev. X* **6**(3) 031021
[11] Rhodes D, Schönemann R, Aryal N, Zhou Q, Zhang Q R *et al.* 2017 *Phys. Rev. B* **96**(16) 165134
[12] Aryal N and Manousakis E 2019 *Phys. Rev. B* **99**(3) 035123

Some aspects of a two-pore translocation problem

Aniket Bhattacharya

Department of Physics, University of Central Florida, Orlando, USA

E-mail: Aniket.Bhattacharya@ucf.edu

Abstract. We report Brownian dynamics (BD) simulation results for a coarse-grained (CG) model semi-flexible polymer threading through two nanopores. Particularly we study a "tug-of-war" situation where equal and opposite forces are applied on each pore to avoid folds for the polymer segment in between the pores. We calculate mean first passage times (MFPT) through the left and the right pores and show how the MFPT decays as a function of the off-set voltage between the pores. We present results for several bias voltages and chain stiffness. Our BD simulation results validate recent experimental results and offer avenues to further explore various aspects of multi-pore translocation problem using BD simulation strategies which we believe will provide insights to design new experiments.

1. Introduction

In a series of experiments [1]-[4] where single-stranded DNA (ss-DNA) were allowed to translocate through an α-hemolysin protein pore by applying a voltage across the pore, it was observed that the histogram of the mean first passage time (MFPT) are very different for different ss-DNA. This lead to the idea that characteristics of translocation through a nano-pore can be used for DNA sequencing. Since then polymer translocation (PT) problem [5] has remained an active field for both fundamental and experimental research. The original aim and scope of rapid and cheap sequencing of human genome without amplification has broadened to analyze other important biopolymers. Technological advancement made it possible to prepare synthetic nano-pores including single layer of graphene nanopores, where the pore diameter and width can be controlled. One of the hallmarks of PT problem is the theoretical and simulation studies of the first passage time problem [5] where ordinary and partial Fokker-Planck equations are used [8, 9] to predict the translocation exponent α for a polymer chain of length N characterized by $\langle\tau\rangle = AN^{\alpha}$, originally proposed by Kantor and Kardar [10]. For the driven translocation problem the tension propagation (TP) theory of Sakaue [11] captured the nonequilibrium aspects of the translocation manifested in *cis-trans* asymmetry [12, 13, 14]. Sakaue's TP theory was subsequently translated into a Brownian dynamics tension propagation (BDTP) scheme [15, 16] and it was possible to carry out simulation of very large polymer chains which validated the scaling theory of driven polymer translocation [17]. It was shown that the finite chain length effects are strong, and it is only in the very large chain limit one observes the asymptotic Kantor-Kardar limit, namely, $\langle\tau\rangle = AN^{1+\nu}$, where ν is the Flory exponent [10].

While single-pore translocation paved the way to do new science with nano-pores and nano-channels, recently it has been realized that instead of a single pore, a ssDNA co-captured in a multiple-nano-pore system with automatic feed-back loop to monitor the speed will provide better control on the polymer segments in between the successive nano-pores [18]. It has been

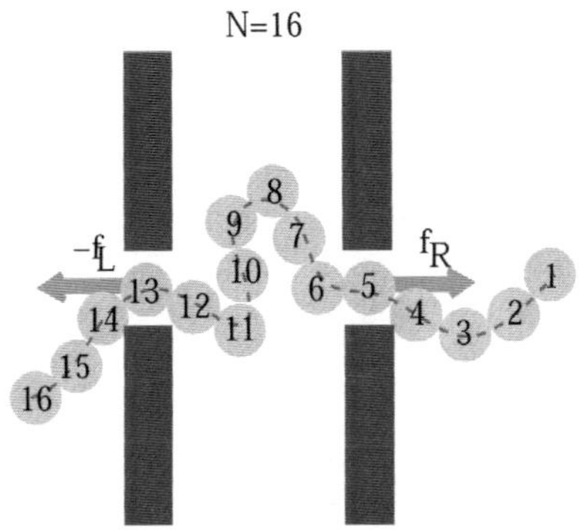

Figure 1. Schematic of BD simulation model for a chain of length $N = 16$. The width of the repulsive wall is 1.2σ and the pore diameter is 2σ so that it can accommodate one particle at a time. External biases are applied in opposite directions as shown.

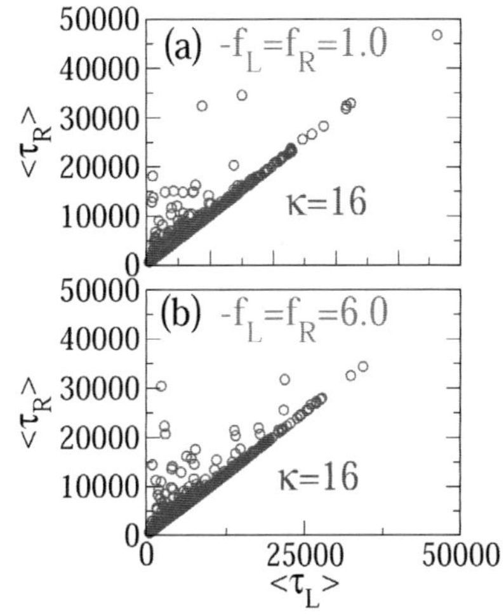

Figure 2. Plot of $\langle \tau_R \rangle$ versus $\langle \tau_L \rangle$ for a tug-of-war situation (a) $-f_L = f_R = 1.0$ and (b) $-f_L = f_R = 6.0$.

recently reported that by applying an equal and opposite bias at the left and right pores of a two-pore system to create a "tug-of-war" situation one can reduce the diffusion of the chain segment and reduce the probability of occurrence of folds which hinders the analyses of the segment [18].

To design such an experiment, where a DNA segment is co-captured in two pores, is a significant challenge. Furthermore, direct observation of physical quantities become very limited due to resolution issues making "seeing" in nano-meter length scales extremely hard. Recently we demonstrated that BD simulation of CG models can provide detail picture of evolution of polymer conformations inside nanochannels [20] which are of considerable help to experimentalists to understand their data with reference to a specific model. In this paper we demonstrate that BD simulation results on a CG model mimicking the two-pore experiment qualitatively captures the experimental results. In the next two sections we briefly describe the model and present preliminary results. We then conclude with a few remark about future work on two-nano-pore systems.

2. The Two-NanoPore Model and BD simulation

We have implemented BD simulation scheme in two dimensions (2D) whose schematics are shown in Fig. 1 for a chain of length $N = 16$ although the simulation is carried out for the chain length $N = 64$. We start with the situation where the polymer segment is co-captured in both the pores. The origin of the coordinate system is chosen at the symmetric location equidistant from the two pores. Initially a straight polymer chain is placed symmetrically with respect to the origin so that the equal straight segments lie at the left of the left pore and right of the right pore. In this paper we kept the distance between the pores $N_0 = 20$ the same, so that at $t = 0$ the beads with indices 23 and 42 are located inside the right and the left pore, beads with indices 1-22 and 43-64 are located at the left and right side of the left and right pore respectively. The beads with indices 24-41 are inside the two pores. The initial straight chain is then equilibrated with BD simulation time about 5 times the Rouse relaxation time $\tau_{eq} \propto N^{1+2\nu}$ keeping the beads located inside each pore clamped. To model the polymer chain, the immobile wall which contains the nanopore, we have used Lennard-Jones (LJ) and anharmonic spring potentials, and a three body bond-bending potential as in Eqns. 1-3, described in detail in our recent publication [20]. The excluded volume interaction between any two monomers is given by a short range Lennard-Jones (LJ) potential.

$$\begin{aligned} U_{\rm LJ}(r) &= 4\epsilon\left[\left(\frac{\sigma}{r}\right)^{12} - \left(\frac{\sigma}{r}\right)^{6}\right] + \epsilon, \text{ for } r \le 2^{1/6}\sigma; \\ &= 0, \text{ for } r > 2^{1/6}\sigma \end{aligned} \tag{1}$$

Here, σ is the effective diameter of a monomer, and ϵ is the strength of the potential. The connectivity between neighboring monomers is modeled as a Finite Extension Nonlinear Elastic (FENE) spring with

$$U_{\rm FENE}(r) = -\frac{1}{2}kR_0^2 \ln\left(1 - r^2/R_0^2\right) \tag{2}$$

where r is the distance between the consecutive monomers, k is the spring constant and R_0 is the maximum allowed separation between connected monomers. The chain stiffness is introduced by adding an angle dependent three body interaction term between successive bonds as

$$U_{\rm bend}(\theta_i) = \kappa\,(1 - \cos\theta_i) \tag{3}$$

Here θ_i is the angle between the bond vectors $\vec{b}_{i-1} = \vec{r}_i - \vec{r}_{i-1}$ and $\vec{b}_i = \vec{r}_{i+1} - \vec{r}_i$, respectively. The parameter $\kappa = k_BT\ell_p$, where ℓ_p is the persistence length of the chain, k_B and T are the Boltzmann constant and the temperature respectively. The values of the parameters and the BD scheme are the same as described in [20].

After the polymer chain is equilibrated with beads inside the left and right pores at clamped position, the chain is allowed to translocate with biases applied at the left and the right pores as shown in Fig. 1. We consider both the cases $-f_L = f_R$ and $|f_L| - |f_R| \neq 0$. The simulation results are averaged over 1000 initial conditions. We have checked in few cases that a choice of initial configuration of the chain as an off-lattice self-avoiding random walk does not affect the final results.

3. Results

We show BD simulation results for a chain length of $N = 64$ for several biases and chain persistence length, and make comparison with the experimental results reported by Reisner and coworkers [18]. First we show the plot of $\langle\tau_R\rangle$ versus $\langle\tau_L\rangle$ in a "tug-of-war" in Fig. 2. The graph shows that most of the points lie on the straight line $\langle\tau_R\rangle = \langle\tau_L\rangle$. The same behavior has been observed by Reisner and coworkers. However, there are scattered points above the line $\langle\tau_R\rangle = \langle\tau_L\rangle$. These points correspond to those events where the chain spends different time in between the pores after finishing translocation through the left pore, and before the final exit through the right pore. We have shown results for two different "tug-of-war" situation; (i) $-f_L = f_R = 1.0$ corresponds to the low bias case when $|f_L|\sigma = |f_R|\sigma \simeq k_BT$ and (ii) the other for higher bias $|f_L|\sigma = |f_R|\sigma \simeq 5k_BT$. In the former case the MFPT is longer resulting the distribution of points to be qualitatively different than the case for $|f_L|\sigma = |f_R|\sigma \simeq 5k_BT$ analyzed further in the context of MFPT in Figs. 3(c) and 4(c). The histograms of the MFPT for the case when the external biases $-f_L = f_r \simeq k_BT$ and $-f_L = f_r \simeq 5k_BT$ are shown in Figs. 3 and 4. The MFPT through the left pore $\langle\tau_L\rangle$ (Figs. 3(a) Figs. 4(a)) and through the right pore (Figs. 3(b) Figs. 4(b)) are qualitatively similar in each case for low and high biases. This is due to the fact that almost 90% of the translocation time is spent before the chain translocates through the left pore (please see the numbers for $\langle\tau_L\rangle$ and $\langle\tau_R\rangle$ in Figs. 3 and 4) when the chain

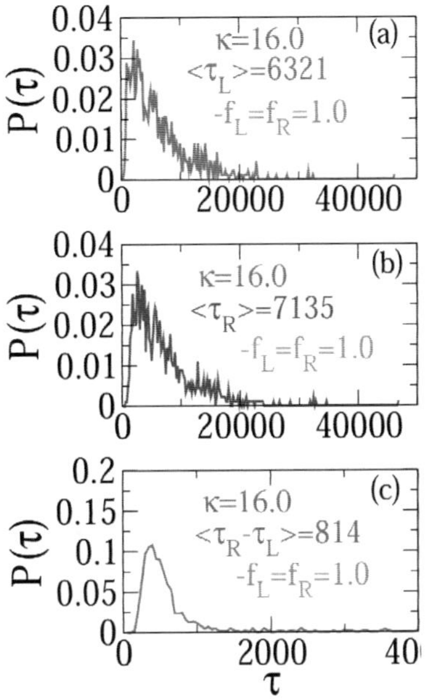

Figure 3. Histogram of the MFPT for $f_L = -1.0$ and $f_R = 1.0$

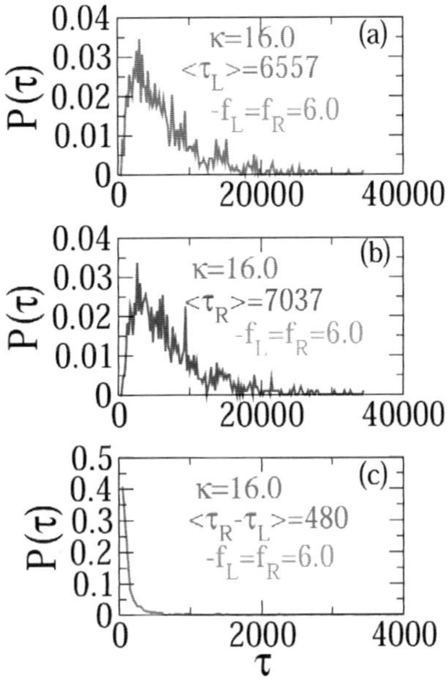

Figure 4. Histogram of the MFPT for $f_L = -6.0$ and $f_R = 6.0$

experiences a tug-of-war situation with net zero force. It is only after the translocation through the left pore the chain experiences the driving force from the right pore only. This explains the qualitative differences in Fig. 3(c) and Fig. 4(c) which corresponds to the scattered points in Fig. 2(a) and 2(b) respectively. The distribution of delay time $|f_L| - |f_R|$ in Fig. 4(c) is narrower than that of in Fig. 3(c), as in the former case the chain experiences a larger net larger force after translocating through the left pore.

Finally, we show the result how $\langle \tau_R \rangle$ depends on the offset bias ($|f_R| - |f_L| > 0$). Evidently one would expect that $\langle \tau_R \rangle$ will be maximum when $\Delta f_{LR} = |f_R| - |f_L| = 0$ and will decay as Δf_{LR} is increased. This is shown in Fig. 5 for two different chain stiffness $\kappa = 16$ and 64. This is in accord with our previous work that a stiffer chain takes longer time to translocate [21]. When we normalize the translocation time by the maximum value in each case, the scaled plots in Fig. 6 collapse onto the same master curve. The general shape of the collapsed curve is the same as observed experimentally [18].

4. Summary and conclusion

In summary, motivated by recent experimental work we have carried out BD simulation to study translocation and eventual escape of a model semi-flexible polymer initially captured by two nanopores. BD simulation on the same model polymer for a single nano-pore translocation have been very successful in explaining a variety of translocation phenomena and played a major role to validate the scaling theories and tension propagation theory of polymer translocation. Therefore, it is no surprise that for the two-pore translocation problem we observe that the the simulation strategy is capable of reproducing the experimental results and thus can provide further details not captured in the experiments. To the best of our knowledge this is the first simulation study of the two-pore translocation problem. But the preliminary results are

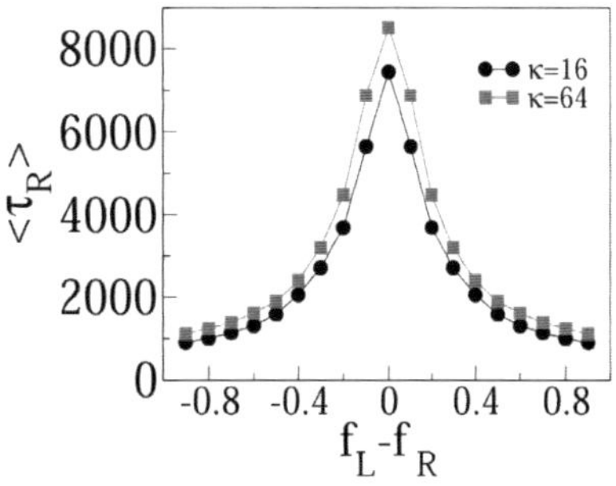

Figure 5. $\langle \tau_R \rangle$ as a function of Δf_{LR} for two values of chain stiffness parameter $\kappa = 16.0$ (black circles) and $\kappa = 64.0$ (red squares) respectively.

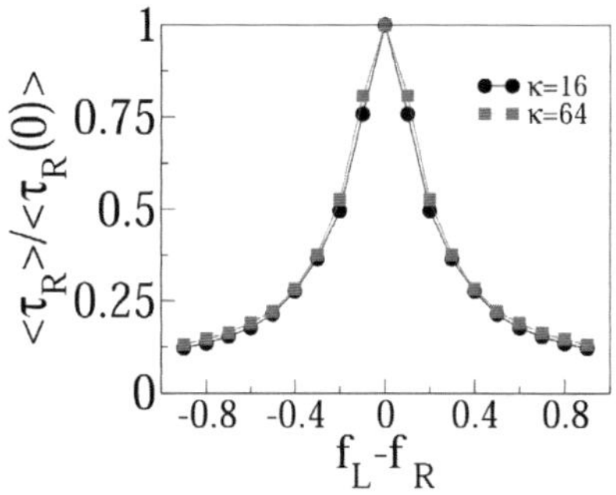

Figure 6. Normalized $\langle \tau_R \rangle$ as a function of Δf_{LR} for two values of chain stiffness parameter $\kappa = 16.0$ (black circles) and $\kappa = 64.0$ (red squares) respectively.

reassuring enough to explore in detail for a larger parameter space for much longer chains which will enable us to watch folds during the translocation process. We also plan to introduce a self-feedback mechanism to slow down the diffusion of the chain which will help to design new experimental strategies to control biomolecules.

5. Acknowledgement

The author thanks Prof. Walter Reisner for various discussions.

[1] J. J. Kasianowicz, E. Brandin, D. Branton and D. W. Deamer, *Proc. Natl. Acad. Sci. U.S.A.* **93**, 13770 (1996).
[2] A. Meller, L. Nivon, E. Brandin, J. A. Golovchenko, and D. Branton, *Proc. Natl. Acad. Sci. U.S.A.* **97**, 1079 (2000).
[3] A. Meller, L. Nivon, and D. Branton, *Phys. Rev. Lett.* **86**, 3435 (2001).
[4] A. Meller and D. Branton, *Electrophoresis* **23**, 2583 (2002).
[5] For recent reviews in the field please see M. Muthukumar *Polymer Translocation* (CRC Press, Boca Raton, 2011); Vladimir V. Palyulin, Tapio Ala-Nissila, and Ralf Metzler, Soft Matter, **10**, 9016 (2014).
[6] W. Sung and P. J. Park, *Phys. Rev. Lett.* **77**, 783 (1996).
[7] M. Muthukumar, *J. Chem. Phys.* **111**, 10371 (1999).
[8] J. L. A. Dubbeldam, A. Milchev, V.G. Rostiashvili, and T.A. Vilgis, Phys. Rev. E **76**, 010801(R) (2007).
[9] J. L. A. Dubbeldam, A. Milchev, V.G. Rostiashvili, and T.A. Vilgis, Europhys. Lett. **79**, 18002 (2007).
[10] Y. Kantor and M. Kardar, *Phys. Rev. E* **69** 021806 (2004).
[11] Sakaue, T *Phys. Rev. E* **76** 021803 (2007).
[12] Lehtola V. V. Kaski K. and Linna R. P *Phys. Rev. E***82** 031908 (2009).
[13] Bhattacharya A and Binder K *Phys. Rev. E***81** 041804 (2010).
[14] A. Bhattacharya, Computer Simulation Studies in Condensed Matter Physics XXII, Eds. D. P. Landau, S. P. Lewis, and H. B. Schuttler, Elsevier, Physics Proceedia **3**, 1411 (2010).
[15] Ikonen T , Bhattacharya A., Ala-Nissila T. and Sung W. *J. Chem. Phys.***137** 085101 (2012).
[16] T. Ikonen, A. Bhattacharya, T. Ala-Nissila and W. Sung, Phys. Rev. E **85**, 051803 (2012).
[17] Ikonen T, Bhattacharya A, Ala-Nissila T, and Sung W *EPL***103** 38001 (2013).
[18] *Controlling DNA Tug-of-War in a Dual Nanopore Device*, Xu Liu, Yuning Zhang, Roland Nagel, Walter Reisner, William B. Dunbar, arXiv:1811.11105
[19] Grest, G.S.; Kremer, K. Molecular dynamics simulation for polymers in the presence of a heat bath. Phys. Rev. A **1986**, *33*, 3628(R) (1986).
[20] Bernier S, Huang A, Reisner W, and Bhattacharya A *Macromolecules* **51** 4012 (2018).
[21] Adhikari R. and Bhattacharya A. *J. Chem. Phys.* **138** 204909 (2013).

Magnetic properties of CrFeCoNi based high entory alloy

Junqi Yin and Markus Eisenbach
National Center for Computational Sciences
Oak Ridge National Laboratory, Oak Ridge, TN, USA

Makrus Daene and G. Malcolm Stocks
Materials Science and Technology Division
Oak Ridge National Laboratory, Oak Ridge, TN, USA

Abstract. Monte Carlo simulations are performed on three high entropy alloys: $Cr_{0.25}Fe_{0.25}Co_{0.25}Ni_{0.25}$, $Cr_{0.2}Fe_{0.2}Co_{0.2}Ni_{0.2}Pd_{0.2}$, and $Cr_{0.2}Mn_{0.2}Fe_{0.2}Co_{0.2}Ni_{0.2}$, with exchange interactions extracted from The ab initio Korringa-Kohn-Rostoker method combined with the coherent potential approximation calculations. Using finite size scaling analyses, we estimate the magnetic phase transition temperature for the four component alloy to be 108(2) K, and although the individual critical exponents are different from 3D Heisenberg universality class, the reduced exponent follows Suzuki weak universality. With the additional Palladium component, the transition temperature elevates to about 200 K. In contrast, we find no magnetic order for the five component alloy with Manganese at any finite temperatures.

1. Introduction

The high entropy alloy (HEA) [1, 2] has increasingly drawn interests in materials science, since it opens opportunities to create novel alloys with better properties than traditional ones. However, our understanding of HEAs is still limited, especially about magnetic properties, which is important for applications requiring magnetic materials at elevated temperatures. Recent experimental studies [3, 4] measured Curie temperatures of several HEAs using CrFeCoNi as base ingredients. It is reported that CrFeCoNi is paramagnetic at finite temperatures and CrFeCoNiPd exhibits a transition temperature at about 200 K [4]. A recent study [5] also shows a transition from the paramagnetic state for CrFeCoNi to the ferromagnetic state for CrFeCoNiAl at room temperature.

In this paper, we perform Monte Carlo simulations to study magnetic properties of three different CrFeCoNi based HEAs with interaction parameters calculated from The ab initio Korringa-Kohn-Rostoker method combined with the coherent potential approximation (KKR-CPA) calculations. In contrast to Ref.[4] and [5], we observe a magnetic phase transition at a low temperature and find that the critical behavior obeys the Suzuki weak universality class.

In Sec. 2, the model, relevant methods and analysis techniques are reviewed. Our results are presented in Sec. 3 $\sim$ 5, and we conclude in Sec.6.

2. Model and methods

2.1. Model

For all three HEAs: CrFeCoNi, CrFeCoNiPd, and CrMnFeCoNi, we only consider the case that each species has the same concentration and is randomly distributed on a FCC lattice. With exchange interactions extracted from KKR-CPA calculations, a Heisenberg model with site-dependent interactions is employed to describe the alloy systems, which is justified by the fact that no significant longitudinal spin fluctuation is observed [6] for similar system. The Hamiltonian is defined as follows,

$$\mathcal{H} = \sum_{<i,j>} J_{ij} S_i \cdot S_j, \tag{1}$$

where S_i is a unit vector and J_{ij} is the exchange interaction between the spin i and the spin j, the value of which is determined by the species of atoms at corresponding sites and the shell (distance) with respect to each other, as shown in Fig. 1. For the first shell (i.e. the nearest-neighbor) interactions, any pairs involving Cr or Mn are antiferromagnetic except for Mn-Ni, while other components interact ferromagnetically. As the shell number increases, the J value oscillates around zero and its magnitude rapidly decreases.

2.2. Wang-Landau method

Since there are competing interactions for different shells, the more shells included, the more "glassy" the behavior becomes (i.e. the more barriers in the free energy landscape). To study the ground state structures, Wang-Landau method [7] is applied to drive the system randomly through the energy space. It is realized by updating the system according to the probability inversely proportional to the density of states.

$$W(X|X') = min[1, \frac{g\,(E(X'))}{g\,(E(X))}], \tag{2}$$

where X is the system configuration.

Given an initial guess (e.g. constant), the density of states is updated iteratively. Once it converges, we will have a flat energy histogram which indicates a random walk in energy sampling. In practice, many random walkers [8] are used to speed up the ground state searching. Each walker has its own replica of the system, and they update simultaneously. Information is exchanged only by sharing a common density of state, therefore it is straightforward for the parallel implementation.

2.3. Parallel tempering and cluster update methods

Parallel tempering Monte Carlo simulations [9, 10] are employed to obtain the thermodynamics of the systems. A swap trial is performed with a Metropolis-like probability which satisfies the detailed balance condition. The transition probability from a configuration X_m simulated at temperature T_m to a configuration X_n simulated at temperature T_n would be

$$W(X_m, T_m|X_n, T_n) = min\left[1, exp(-\triangle)\right], \tag{3}$$

$$\triangle = (1/k_BT_n - 1/k_BT_m)(\mathcal{H}_m - \mathcal{H}_n). \tag{4}$$

Since for all three HEAs we considered the specific heat turns out to be non-divergent at the transition point, a simulation temperature set in geometric series works optimally [11].

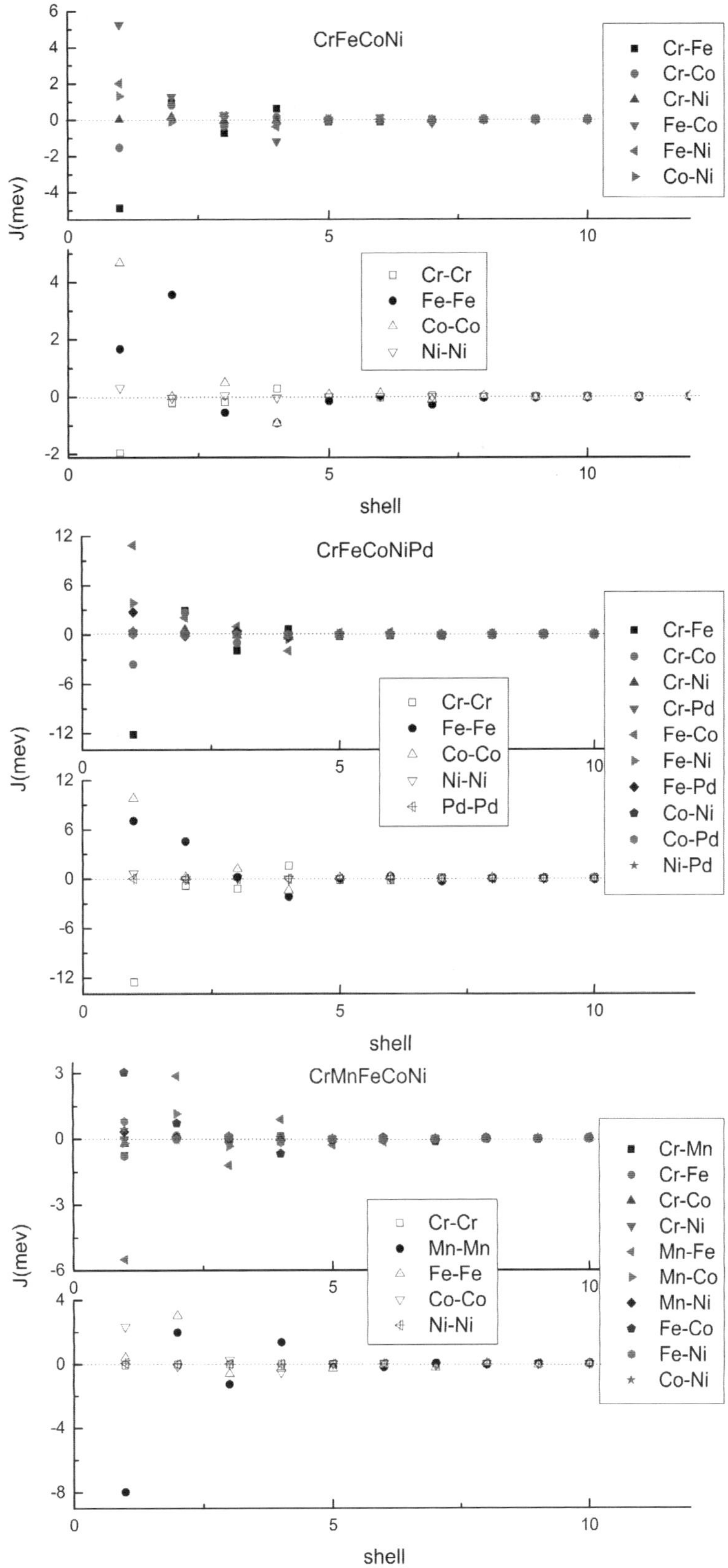

Figure 1. Pair exchange parameters calc[illegible] from KKR-CPA method for CrFeCoNi, CrFeCoNiPd, and CrMnFeCoNi.

To counter the effect of the critical slowing down near the phase transition, we also apply the Wolff cluster update algorithm [12]. Given a randomly chosen direction $\hat{n}$, spins are iteratively added to the flipping cluster with probability

$$p = 1 - exp\left\{min\left[0, \frac{2J_{ij}}{k_B T}(\hat{n}\cdot S_i)(\hat{n}\cdot S_j)\right]\right\}, \tag{5}$$

and then the cluster of spins is flipped with respect to the hyperplane orthogonal to $\hat{n}$.

Therefore, a typical Monte Carlo step consists of a single spin flip trial for every spin, a cluster update for the whole system, and a replica swapping trial between neighboring temperatures. For the data shown below, error bars are estimated from 100 different initial configurations of each alloy.

2.4. Finite size scaling

To study the critical behavior, we perform finite-size scaling analyses [13] near the transition point. The observable of interest are specific heat c, magnetization M, susceptibility χ, and fourth order cumulant U, which are defined as follows,

$$c = \frac{1}{Nk_B T^2}\left[< (E)^2 > - < E >^2\right], N = L^3 \tag{6}$$

$$M = \frac{1}{N}\left[\sqrt{(\sum_i S_i^x)^2 + (\sum_i S_i^y)^2 + (\sum_i S_i^z)^2}\right] \tag{7}$$

$$\chi = \frac{N}{T}\left[< (M)^2 > - < M >^2\right] \tag{8}$$

$$U = 1 - \frac{< (M)^4 >}{3 < (M)^2 >^2} \tag{9}$$

We also investigate the magnetization for each component specie with the similar definition as Eq. 7. The difference is that the summation runs over one specified type of atoms. The component χ and U are defined accordingly.

Since the maximum slope of the fourth-order cumulant U follows

$$(\frac{dU}{dK})_{max} = a' L^{\frac{1}{\nu}}(1 + b' L^{-\omega}), \tag{10}$$

where $K = \frac{1}{T}$, the correlation length exponentν can be estimated directly.

With the exponent ν and critical temperature T_c at hand, the critical exponent β and γ can be extracted from the data collapsing of the finite-size scaling forms,

$$M = L^{-\frac{\beta}{\nu}}\overline{X}(tL^{\frac{1}{\nu}}) \tag{11}$$

$$\chi T = L^{\frac{\gamma}{\nu}}\overline{Y}(t' L^{\frac{1}{\nu}}) \tag{12}$$

where $t = |1 - \frac{T}{T_c}|$, $t' = |1 - \frac{T_c}{T}|$ and $\overline{X}$ and $\overline{Y}$ are universal functions whose analytical forms are not known.

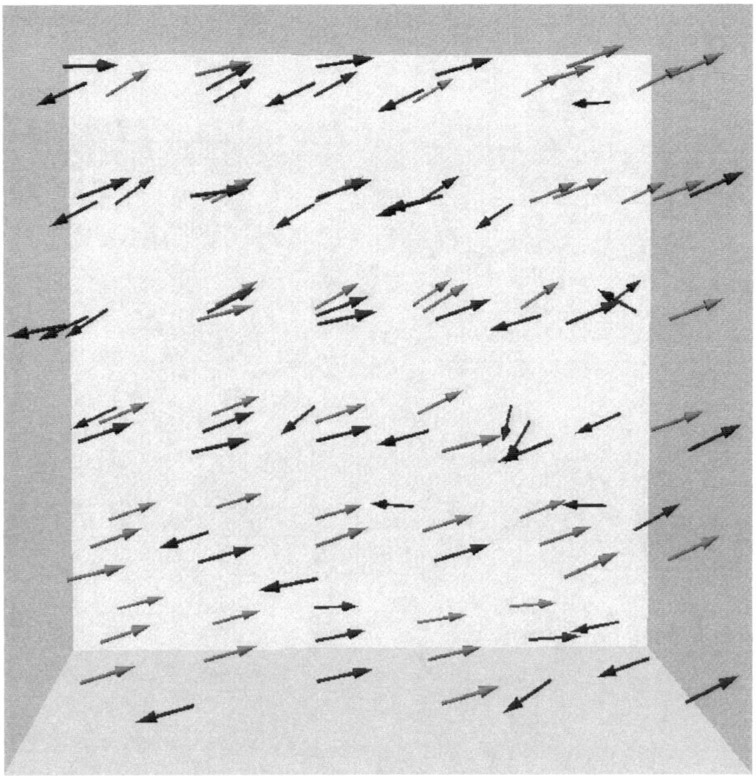

Figure 2. Snapshot of the ground state configuration for CrFeCoNi with $N = 1000$ and up to 8 shells. Red arrows denote Co; yellow arrows denote Fe; light blue arrows denote Ni; dark blue arrows denote Cr.

3. Results: $Cr_{0.25}Fe_{0.25}Co_{0.25}Ni_{0.25}$

The ground state searched by the Wang-Landau random walk is ferromagnetic for up to 20 shells included in the model. Most of the spins are collinearly aligned with Cr spins pointing the opposite direction of the other three types of spins. But some Cr spins form locally non-collinear structures whose directions are almost perpendicular to the overall magnetic direction. A snapshot of the ground state spin configuration is plotted in Fig. 2. Therefore, a magnetic phase transition is expected at some finite temperature for this four component alloy. Due to the computational cost resulting from not only linearly increased neighbors of interactions but also the "glassy" slow dynamics, the thermodynamics data in this section are collected for the model with up to 8 shells.

In Fig. 3, we show the specific heat and susceptibility for six different system sizes. A non-diverging specific heat indicates a negative α exponent, which is similar to the case for the classical Heisenberg model and bulk BCC Iron. A clear signal for the continuous transition is seen from the susceptibility data. The transition temperature is estimated to be 108(2) K according to the fourth order cumulant of the magnetization. Considering the susceptibility for each component, we find that they show more or less the same peak position. While M_{Co} and M_{Ni} has the largest and smallest variance, respectively, the overall susceptibility behaves like that of Iron.

As for the critical behavior, we evaluate the exponent β and γ following the Eq. 11 and Eq. 12,

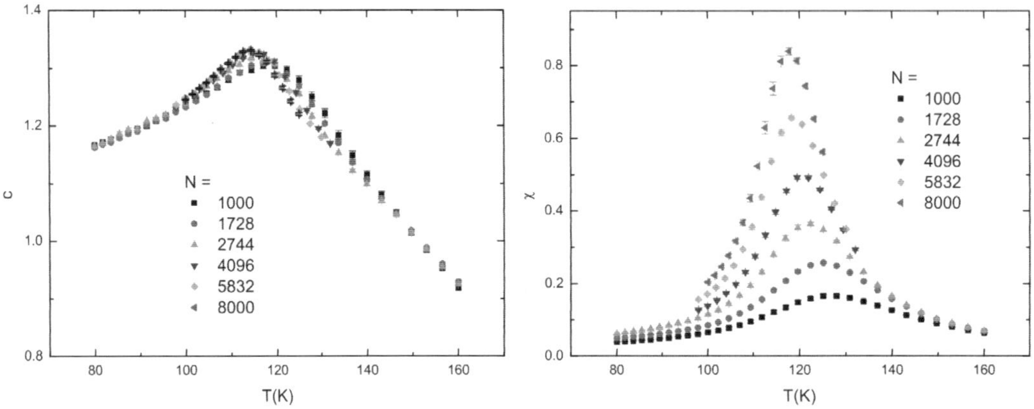

Figure 3. Specific heat and susceptibility plots for CrFeCoNi.

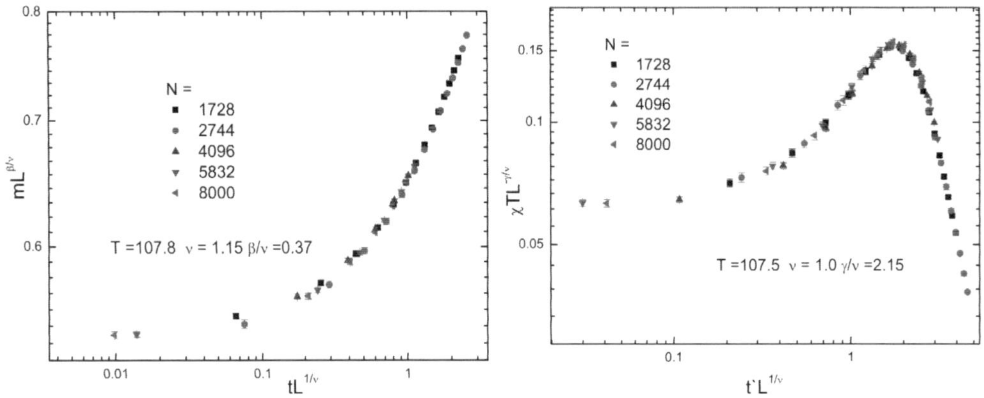

Figure 4. Finite size scaling of the magnetization and its susceptibility, respectively.

respectively. The corresponding data collapsing plots are shown in Fig. 4. Both exponents are larger than those of 3D Heisenberg Universality class, but the reduced exponent $\gamma/\nu = 2.1(2)$ is close to the Universality value 1.96(1). Many systems with competing interactions [14] or quenched disorder [15] present similar behavior which is an indication of Suzuki weak Universality.

4. Results: $Cr_{0.2}Fe_{0.2}Co_{0.2}Ni_{0.2}Pd_{0.2}$

To obtain a similar alloy but with a higher transition temperature, one can add another component that strengthens the ferromagnetic interaction. Pd-Pd has a weak ferromagnetic coupling. See Fig. 1. The ground state of CrFeCoNiPd is similar to CrFeCoNi with Pd spins pointing to the same direction as other ferromagnetic components and Cr spins partially non-collinearly aligned, as shown in Fig. 5.

In Fig. 6, The magnetization and its fourth order cumulant are plotted for the individual component and different lattice sizes, respectively. The critical temperature is estimated to be 339(2) K which is over 200 degree higher than that of the corresponding four component alloy. For the original four components, each shows similar behavior as before, i.e., similar shape of

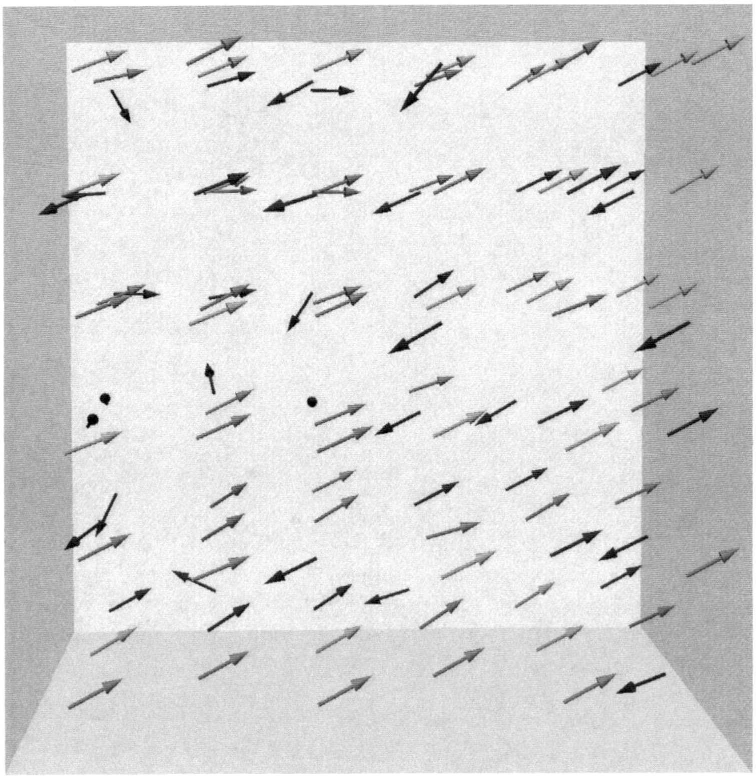

Figure 5. Snapshot of the ground state configuration for $CrFeCoNiPd$ with $N = 1000$ and up to 8 shells. Red arrows denote Co; yellow arrows denote Fe; light blue arrows denote Ni; dark blue arrows denote Cr; silver arrows denote Pd.

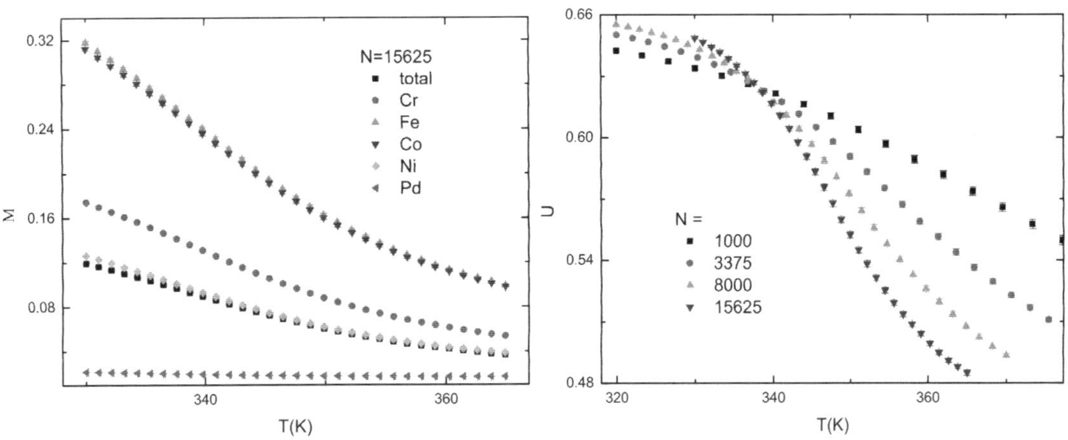

Figure 6. Magnetization and its 4th order cumulant for $CrFeCoNiPd$.

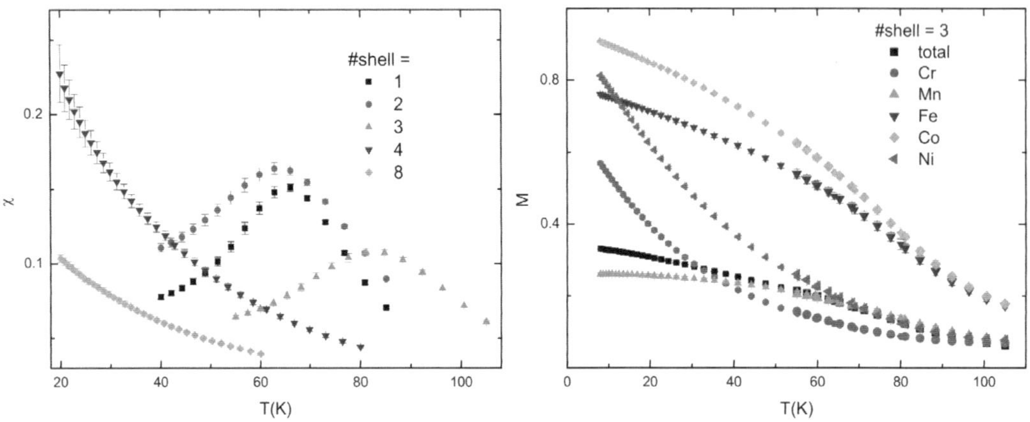

Figure 7. Magnetization and its susceptibility for $CrMnFeCoNi$.

magnetization curve and similar individual "T_c". But spins on Pd becomes disorder at finite temperatures.

From the finite size scaling of susceptibility (Eq. 12) and fourth order cumulant (Eq. 10), we estimate $\nu = 0.85(2)$ and $\gamma = 1.68(4)$ which is significantly larger than that of the 3D classical Heisenberg Universality, respectively. However, $\gamma/\nu = 1.98(3)$ agrees with Universality value 1.96(1) within the error bar. It seems Suzuki week Universality holds.

5. Results: $Cr_{0.2}Mn_{0.2}Fe_{0.2}Co_{0.2}Ni_{0.2}$

Another interesting topic is to investigate the effect of adding an antiferromagnetic component to CrFeCoNi. As shown in Fig. 1, Mn-Mn interaction is strongly antiferromagnetic. With the additional Manganese atom, the resulting properties change dramatically. There exists no overall magnetization at finite temperatures if enough shells (more than 3) are included in the model. With only the first three shells, we find a non-collinear ground state. In Fig. 7(a), the susceptibility for models with different numbers of interaction shells are plotted. Peaks at finite temperatures can be seen for up to 3 shells. By examining the magnetization for the individual component, the contribution to the variance mostly comes from Fe and Co. See Fig. 7(b). This behavior is different from the four component alloy where they all present similar shapes. Essentially, the spins on Cr and Ni atoms become disorder with respect to its specie at finite temperatures. As more shells included, even M_{Fe} and M_{Co} disappear when the temperature is higher than zero.

6. Conclusion

With exchange interactions calculated from KKR-CPA method, We have studied thermodynamic properties of three different high entropy alloys: CrFeCoNi, CrFeCoNiPd, and CrMnFeCoNi. There exists a magnetic phase transition for CrFeCoNi at 108(2) K and the critical behavior follows Suzuki weak universality. By adding Palladium to the four component alloy, we obtain a much higher transition temperature at 339(2) K. The critical behaviors of both alloys seem to follow Suzuki weak Universality. On the other hand, CrMnFeCoNi shows no magnetization at finite temperatures.

Acknowledgments
Acknowledgments This work was sponsored by the U.S. Department of Energy, Office of Science, Basic Energy Sciences, Materials Science and Engineering Division. This research used resources of Oak Ridge National Laboratory's Compute and Data Environment for Science (CADES) and the Oak Ridge Leadership Computing Facility, which is supported by the Office of Science of the U.S. Department of Energy under Contract No. DE-AC05-00OR22725.

[1] J. W. Yeh, S. K. Chen, S. J. Lin, J. Y. Gan, T. S. Chin, T. T. Shun, C. H. Tsai, and S. Y. Chang, Adv. Eng. Mater. **6**, 299 (2004).
[2] J. W. Yeh, Y. L. Chen, S. J. Lin, S. K. Chen, Mater. Sci. Forum **560**, 1 (2007).
[3] K. B. Zhang, Z. Y. Fu, J. Y. Zhang, J. Shi, W. M. Wang, H. Wang, Y. C. Wang, and Q. J. Zhang, J. Alloys Compd. **502**, 295 (2010).
[4] M. S. Lucas, L. Mauger, J. A. Muñoz, Y. Xiao, A. O. Sheets, S. L. Semiathin, J. Horwath, and Z. Turgut, J. Appl. Phys. **109**, 07E307 (2011).
[5] S. Huang, W. Li, X. li, and S. Schönecker, Materials design **103** 71 (2016).
[6] S. Mu, J. Yin, G. D. Samolyuk, S. Wimmer, Z. Pei, M. Eisenbach, S. Mankovsky, H. Ebert, G. M. Stocks, Phys. Rev. Materials **3**, 014411 (2019).
[7] F. G. Wang and D. P. Landau, Phys. Rev. Lett. **86** 2050 (2001); Phys. Rev. E**64** 056101 (2001);
[8] L. Zhan, Comp. Phys. Comm. 179 (2008) 339; M. Eisenbach, C.-G. Zhou, D. M. Nicholson, G. Brown, J. Larkin and T. C. Schulthess, SC, Portland, Oregon, USA, November 14-20. ACM, New York (2009); J. Yin and D. P. Landau, J. Chem. Phys. **134**, 074501 (2011); Comp. Phys. Comm. **183**, 1568 (2012).
[9] R. H. Swendsen and J. S. Wang, Phys. Rev. Lett. **57**, 2607(1986).
[10] K. Hukushima and K. Nemoto, J. Phys. Soc. Japan **65**, 1604 (1996).
[11] D.A. Kofke, J. Chem. Phys. **120**, 10852 (2004).
[12] U. Wolff, Phys. Rev. Lett. **62**, 361 (1989).
[13] See, for example, *Finite Size Scaling and Numerical Simulation of Statistical Systems*, edited by V. Privman (World Scientific, Singapore, 1990); D. P. Landau and K. Binder, *A Guide to Monte Carlo Simulation in Statistical Physics* (Cambridge University Press, Cambridge, U.K., 2000).
[14] J. Yin and D. P. Landau, Phys. Rev. E **80**, 05117 (2009); Phys. Rev. E **81**, 031121 (2010).
[15] J. Yin, B. Zheng, and S. Trimper, Phys. Rev. E **70**, 056134 (2004); Phys. Rev. E **72**, 036122 (2005).

A 64 bit quantum dragon data-set for machine learning

M.A. Novotny[*,†], **Yaroslav Koshka**[*,‡], **G. Inkoom**[†], **Vivek Dixit**[*]

[*] Center for Computational Sciences, High Performance Computing Collaboratory, Mississippi State University, Mississippi State, Mississippi 39762-9627, USA
[†] Department of Physics and Astronomy, Mississippi State University, Mississippi State, Mississippi 39762-5167, USA
[‡] Department of Electrical and Computer Engineering, Mississippi State University, Mississippi State, Mississippi 39762, USA

E-mail: man40@msstate.edu

Abstract. Design and examples of a sixty-four bit quantum dragon data-set are presented. A quantum dragon is a tight-binding model for a strongly disordered nanodevice, but when connected to appropriate semi-infinite leads has complete electron transmission for a finite interval of energies. The labeled data-set contains records which are quantum dragons, which are not quantum dragons, and which are indeterminate. The quantum dragon data-set is designed to be difficult for trained humans and machines to label a nanodevice with regard to its quantum dragon property. The 64 bit record length allows the data-set to be utilized in restricted Boltzmann machines which fit well onto the D-Wave 2000Q quantum annealer architecture.

1. Introduction

Machine learning is making inroads into many aspects of modern research and modern society [1], from social media algorithms to autonomous vehicle applications. Machine learning is also making inroads as an application used in computational [2, 3] and experimental science [4]. One type of machine learning methodology is restricted Boltzmann machines (RBM) [5]. The possibility of using quantum computers in machine learning has a recent review [6]. The application of quantum annealing machines to train and utilize RBM methods [7, 8] is well established. Here we note an RBM can be embedded onto the $K_{4,4}$ Chimera architecture of a D-Wave machine [9, 10, 11], and the current D-Wave 2000Q can embed an RBM with 64 hidden and 64 visible units (assuming no missing qubits). In order to test classical and quantum machine learning methodologies, a data set based on quantum, rather than classical, principles may be important.

Thus data-sets with 64 bit records which are based on quantum principles are useful to test current D-Wave architectures as applied to machine learning. The goal of this paper is to introduce the design of a labeled 64 bit data-set which is difficult for humans and RBMs embeddable on a D-Wave 2000Q to generalize from a training set to distinguish underlying patterns which are not in the training set. Because of the small record size, construction of such a data-set is not an easy task.

2. Quantum Dragon Nanodevices

The quantum dragon data-set (QDDS) we introduce is based on electron transport through nanodevices which have strong disorder, but when connected to appropriate semi-infinite leads have complete electron transmission, $\mathcal{T}(E) = 1$ over a finite interval of energies. Quantum dragon devices with cylindrical symmetry were introduced in 2014 [12], including Bethe lattices and single-walled carbon nanotubes. Here we construct our QDDS based on a nanodevice with an underlying square lattice [13]. We use both open boundary conditions of two types. Free boundary conditions are denoted BC00 [14], and boundary conditions modified at the boundary, denoted BC-- [13]. Quantum dragon devices are found by performing an exact mapping for the transmission onto a one dimensional device with the same $\mathcal{T}(E)$. This can be accomplished when every slice j of the device is represented by an $m \times m$ intra-slice matrix $\mathbf{A}_j$, and all slices have a common eigenvector $\vec{v}_{\rm dragon}$ with associated eigenvalue λ_j. Table 1 shows the form for $\mathbf{A}_j$ and the common eigenvector $\vec{v}_{\rm dragon}$ for both boundary conditions in our QDDS.

Table 1. The two boundary conditions utilized for 64 bit quantum dragon data-sets. Within the nanodevice the on site energies are $\epsilon_{i,j}$ and the intra-slice hopping strengths are $t_{i,j;i+1,j}$.

BC	$\vec{v}_{\rm dragon}$	$\mathbf{A}_j$
BC00	$\frac{1}{\sqrt{2}}\begin{pmatrix} \sin\left(\frac{\pi}{4}\right) \\ \sin\left(\frac{\pi}{2}\right) \\ \sin\left(\frac{3\pi}{4}\right) \end{pmatrix} = \frac{1}{2}\begin{pmatrix} 1 \\ \sqrt{2} \\ 1 \end{pmatrix}$	$\begin{pmatrix} \epsilon_{1,j} & -t_{1,j;2,j} & 0 \\ -t_{1,j;2,j} & \epsilon_{2,j} & -t_{2,j;3,j} \\ 0 & -t_{2,j;3,j} & \epsilon_{3,j} \end{pmatrix}$
BC--	$\frac{1}{\sqrt{3}}\begin{pmatrix} 1 \\ 1 \\ 1 \end{pmatrix}$	$\begin{pmatrix} \epsilon_{1,j} - t_{1,j;2,j} & -t_{1,j;2,j} & 0 \\ -t_{1,j;2,j} & \epsilon_{2,j} & -t_{2,j;3,j} \\ 0 & -t_{2,j;3,j} & \epsilon_{3,j} - t_{2,j;3,j} \end{pmatrix}$

For our 64 bit QDDS we utilize a device with $\ell = 4$ slices, each with $m = 3$ atoms in the slice, as shown in Figs. 1 and 2. The hopping strength of the semi-infinite lead wires is chosen as the unit of energy. (Only two atoms in each semi-infinite lead wire are shown, as green spheres.) All inter-slice hopping strengths are also set to unity. The on site energy of the atoms in the lead wires is either chosen as the zero (setting the zero for energy), or if all slices have $\vec{v}_{\rm dragon}$ as a common eigenvector and all the associated eigenvalues $\lambda_j = \lambda_{\rm dragon}$ are the same, the on site lead energies are all set to $\lambda_{\rm dragon}$. For zero lead on site energy, one has propagation of electrons for energies in the interval $-2 \leq E \leq 2$. In the figures, the radii of the red spheres are proportional to the on site energies $\epsilon_{i,j}$, and the radii of the black (vertical) cylinders are proportional to the intra-slice hopping strengths $t_{i,j;i+1,j}$. The green spheres are the lead atoms, and the cyan horizontal bonds all have strength unity.

We choose $n_\# = 3$ bits to represent both the intra-slice hopping strengths and the on site energies of the device. We take both the on site energies $\epsilon_{i,j}$ and intra-slice hopping strengths $t_{i,j;i',j}$ to be non-negative, and have the allowed values $\{0, 1, 2, 3, 4, 5, 6, 7\}$ for $n_\# = 3$. Thus in each of the $\ell = 4$ slices there are $2m - 1 = 5$ parameters, represented by $(2m-1)n_\# = 15$ bits. Our $\ell = 4$ devices thus are represented in 60 bits.

We use two additional bits to label the type of boundary condition, BC00 or BC--, leaving an additional bit to label other boundary conditions [13]. Two bits label the record. If the record is known to represent a quantum dragon, the nanodevice has $\mathcal{T}(E) = 1$ for $-2 \leq E \leq 2$, and is coded as 01. If the record representing the device is mappable (*i.e.* all $\mathbf{A}_j$ share a common eigenvector), but the eigenvectors have different eigenvalues for different j, extending particular values to infinite ℓ would lead to transmission $\mathcal{T}(E) \ll 1$ due to Anderson localization, and hence the device is not a quantum dragon, and is coded as 10. If the mapping is not possible, the value for $\mathcal{T}(E)$ is such there may be some range of energies which have complete transmission.

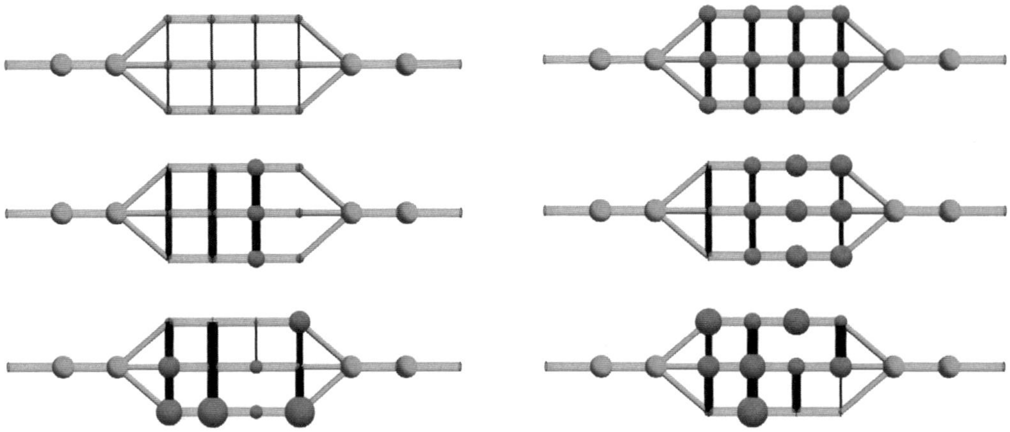

Figure 1. Data set representatives for $\ell = 4$ and $m = 3$ lattice, attached to leads with boundary conditions BC00. The top two are quantum dragons, the middle two are mappable but not quantum dragons, and the last two are undecided.

In this case the device cannot be classified, and it is coded randomly as 00 or 11.

3. Quantum Dragon Data-Sets

We thus have 64 bit records in our QDDS, each representing a nanodevice together with its labels. For our 64 bit QDDS it is possible to calculate the number of possible records, the number of records which correspond to mappable devices, and the number of records corresponding to quantum dragons. These are shown in Table 2.

Table 2. The number of possible different nanodevices for quantum dragon data-sets for an ℓ=4 and m=3 device with $n_{\#} = 3$.

Type	label	# config	# mappable	# dragons
BC00	00	$2^{60} = 1.15 \times 10^{18}$	$2^{24} = 1.68 \times 10^{7}$	$2^{6} = 64$
BC--	01	$(8 \times 36^{2})^{4} = 1.16 \times 10^{16}$	$104^{4} = 1.17 \times 10^{8}$	91208

For BC00 there are $2^{5\times4\times3} = 2^{60}$ different configurations, since there are $2m - 1 = 5$ tight binding parameters per slice, $\ell = 4$ slices, and $n_{\#} = 3$. Of these $2^{2\times4\times3} = 2^{24}$ are known to be mappable, since there are $\ell = 4$ slices and within each slice j one needs $t_{1,j;2,j} = t_{2,j;3,j}$ for the hopping energy values and $\epsilon_{1,j} = \epsilon_{2,j} = \epsilon_{3,j}$ on site energy values. Of these mappable configurations, $2^{2\times3} = 64$ are quantum dragons since in all slices all $t_{i,j;i+1,j}$ values must be the same and all $\epsilon_{i,j}$ values must be the same. See Fig. 1 for representative configurations for BC00 boundary conditions. The BC00 dragon set is relatively easy for humans to determine whether or not the device is a quantum dragon, since there is a high degree of symmetry because every slice must be identical.

For BC-- one requires $t_{1,j;2,j} \leq \epsilon_{1,j}$ and $t_{m-1,j;m,j} \leq \epsilon_{m,j}$ so both the tight binding parameters and the elements of the $\mathbf{A}_j$ matrices can be represented in $n_{\#}$ digits. For $n_{\#} = 3$ the calculation $\sum_{\epsilon=0}^{7} \sum_{t=\epsilon}^{7} 1 = 36$ gives the number of possible tight binding parameters for the boundary atoms and

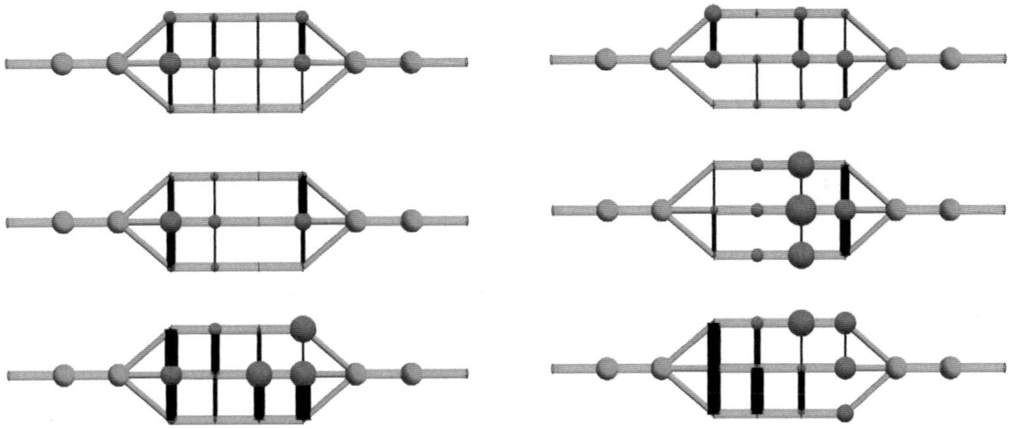

Figure 2. Data set representatives for $\ell = 4$ and $m = 3$ lattice, attached to leads with boundary conditions BC$--$. The top two are quantum dragons, the middle two are mappable but not quantum dragons, and the last two are undecided.

their attached intra-slice hopping parameters, compared to the $2^{2n_\#} = 64$ representable values. Therefore since one requires for putting into bits $\epsilon_{1,j} \geq t_{1,j;2,j}$ and $\epsilon_{m,j} \geq t_{m-1,j;m,j}$ while all other $\epsilon_{i,j}$ can take $2^{n_\#}$ values, for $m = 3$ there are $(8 \times 36^2)^4$ possible records. A Mathematica program was written to test each of the $8 \times 36^2 = 10,368$ allowed values for $\mathbf{A}_j$, and obtained 104 which had the appropriate vector $\vec{v}_{\rm dragon}$ as an eigenvector with eigenvalue λ_j. Therefore, there are 104^4 mappable configurations. Of these mappable configurations, the program gives for BC-- and $\ell = 4$, after taking into account degeneracies, a total of 91208 which are quantum dragons. See Fig. 2 for representative configurations for BC-- boundary conditions. As seen in Fig. 2, the BC-- QDDS is difficult for humans to utilize to identify quantum dragons, since the quantum dragon devices have no easily recognizable symmetry.

4. Conclusion and Discussion

A data-set with 64 bits was designed, because such a data-set can test a restricted Boltzmann machine which can be fit easily onto a D-Wave 2000Q quantum annealer [9, 10, 11]. The data set is built on a $m \times \ell$ square lattice nanodevice in the tight binding representation, with boundary conditions BC00 or BC--, and hence has $\ell(2m-1)$ tight binding parameters. Each tight binding parameter is taken to be non-negative, and embedded in $n_\#$ bits. For our data-sets we choose $\ell = 4$, $m = 3$, and $n_\# = 3$, giving 60 bits to provide a record for a particular nanodevice. We use two additional bits to label whether or not the device is a quantum dragon or if the answer is unknown, as well as two bits to label the boundary condition used in the nanodevice.

Future data-sets can use different values for ℓ, m, and $n_\#$ if larger restricted Boltzmann machines are embedable on a future quantum annealers. For the BC00 the number of total possible configurations are $\left(2^{(2m-1)n_\#}\right)^\ell$, the number of mappable configurations is $2^{2\ell n_\#}$, and the number of quantum dragon configurations is $2^{2n_\#}$. For BC-- the number of possible configurations is $\left[2^{(2m-5)n_\#} \times \left(\frac{2^{2n_\#}+2^{n_\#}}{2}\right)^2\right]^\ell$, while the number of mappable and quantum dragon configurations can be obtained by writing a program to count the number, including taking into account symmetries.

The quantum dragon data-sets are very quantifiable. It is possible to calculate the total number of records, the number of mappable devices, and the number of quantum dragons. Thus a data set is comprised by choosing at random the three types of devices, each with a prescribed probability. Furthermore, some records can be left out of a training data-set, and left for a test data-set. The quantifiable nature of the data-set enables a measure of how well either a human or a machine learning paradigm can generalize from the training data-set to records which were not used in the training.

Acknowledgments
This material is based on research sponsored by the Air Force Research Laboratory (AFRL) under agreement number FA8750-18-1-0096. The U.S. Government is authorized to reproduce and distribute reprints for Governmental purposes notwithstanding any copyright notation thereon. The views and conclusions contained herein are those of the authors and should not be interpreted as necessarily representing the official policies or endorsements, either expressed or implied, of the Air Force Research Laboratory (AFRL) or the U.S. Government.

References

[1] Mohri, A., Rostamizadeh, A., & Talwakar, A. 2018 *Foundations of Machine Learning*, 2nd edition (Cambridge, MA, USA; MIT Press).
[2] Carleo, G. & Troyer, M. Science **355**, 602-606 (2017).
[3] Tarelho, J.P.G., *et al*, Materials Today **21**, 1019-1041 (2018).
[4] Wigley, P.B. *et al.* Scientific Reports **6**, 25890 [6 pages] (2016).
[5] Aggarwal C.C. 2018 *Neural Networks and Deep Learning: A Textbook* (Cham, Switzerland; Springer International Publishing).
[6] Biamonte, J. *et al.* Nature **549**, 195-202 (2017).
[7] Adachi, S.H. & Henderson, M.P. arXiv:1510.06356 (2015).
[8] Rasdi Rere, L.M. Fanany, M.I., & Arymurthy, A.M. Procedia Computer Science **72**, 137-144 (2015).
[9] Koshka, Y., Perera, D., Hall, S., & Novotny, M.A. Proceedings of the International Joint Conference on Neural Networks, 1948-1954 (2016).
[10] Koshka, Y., Perera, D., Hall, S., & Novotny, M.A. Neural Computation **29**, 1815-1837 (2017).
[11] Koshka, Y., Perera, D., Hall, S., & Novotny, M.A. Proceedings of the International Joint Conference on Neural Networks, 8489746 (2018).
[12] Novotny, M.A. Physical Review B **90**, 165103 [14 pages] (2014).
[13] Inkoom, G. Ph.D. dissertation, Mississippi State University (2017).
[14] Inkoom, G. & Novotny, M.A. J. Physics Communications **2**, 115109 [22 pages] (2018).

Power-law distribution found in city-scale traffic flow simulation

Daigo Umemoto[1] **& Nobuyasu Ito**[1,2]
[1]RIKEN Center for Computational Sciences,
7-1-26 Minatojima-minami-machi, Chuo-ku, Kobe, Hyogo 650-0047, Japan
Tel.: +81-78-940-5868
[2]Department of Applied Physics, Graduate School of Engineering, The University of Tokyo
7-3-1 Hongo, Bunkyo-ku, Tokyo 113-8656, Japan
E-mail: daigo.umemoto@riken.jp

Abstract. Origin of a power-law in traffic-volume distribution found in traffic simulations of Kobe city was studied. The traffic distribution which was obtained from a shortest path search with randomized OD (origin-destination) set in Kobe city digital map obeys power-law. The toy model that Cayley tree is embedded in the network is also verified. It is theoretically shown that the traffic distribution with all possible OD set in a Cayley tree obeys power-law like distribution. With randomized OD set, the distribution is diffused from the theoretical point sets. Relationship between these facts and the origin of power-law is discussed.

1. Introduction

Traffic flow model in a simple case, on a single straight lane, has been established[1]. The next topic is to construct an urban scale traffic flow model from the simple traffic model, which is not still well understood in the aspect of the city road network. Attempts to understand the rough behavior of the traffic model with complex city road networks, such as [2, 3], are still limited in number, therefore it is still difficult to design urban traffic flow based on a systematic and analytical model.

Urban scale traffic, which sometimes spontaneously produce traffic congestions, can be understood as a complex system. Such systems, including our society itself, sometimes produce power-law behavior in such as income distribution [4], web accesses [5] and so on. Especially, the former is empirically analogous to traffic flow. There are popular roads such as bypasses or highways which are limited in number are used by large number of vehicles, while many local roads such as small roads in front of individual houses are not so common in the transportation. Therefore, there seems to be inverted proportional relationship between number of roads and traffic popularity. We demonstrated such relationships actually appears and the relationship obeys power law robustly in [6]. Our result also suggested the structure of the road network is one of candidates which causes the power-law distribution.

If the network itself causes the power-law behavior, simple mathematic mechanism such as path search is considered to be a candidate that causes the behavior. In the present paper, we demonstrate a simple shortest path search actually causes the power law and further discuss the origin of the power-law appearance, by introducing artificially created network based on our toy model.

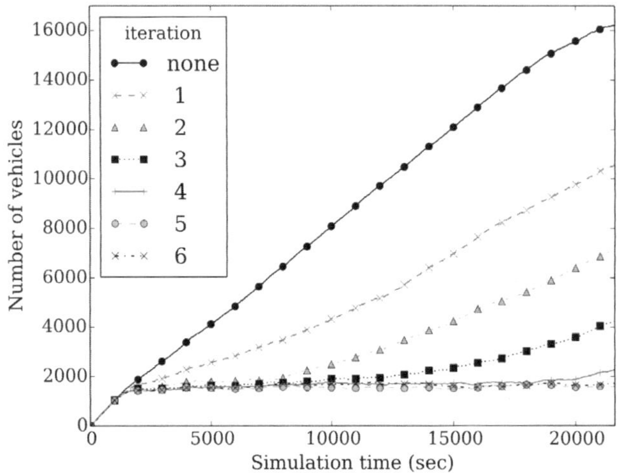

Figure 1. Number of vehicles with increasing traffic path iteration, quoted from [6].

2. power-law behavior in Kobe city agent based traffic simulation

In [6], we performed agent-based traffic simulation using SUMO (Simulation of Urban Mobility) and digitized map created by [3] from Zenrin co. ltd.. Each agent is a vehicle. Each road segment (between a intersection and its nearest next intersection) has its own speed limit which is 100, 60, and 30 km/h. Signals are used with SUMO default settings for simplicity. Simulation duration time is 6 hours and 17,500 vehicles are inserted into the simulation. Refer [6] for more detailed simulation settings. The pairs of origin and destination, hereafter we call OD set, is randomly generated by applying uniform distribution within the entire city.

Once executing a simulation with an above setting, SUMO searches shortest time consuming path for each OD based on road length and speed limit without taking interactions between agents into account, before agents start their drive. As a consequence, severe congestions occur in common roads and non-negligible fraction of vehicles stacks, and does not reach their destinations within 6 hours of simulation time, which are only several kilo-meters apart at maximum. This unrealistic situation occurs because each car never avoid congested roads. To avoid this, SUMO also provides re-route algorithm, which searches path based on not only road length and speed limit, but also the time required to go through each road segment. By this iteration, the simulation becomes more realistic; vehicles avoid severely congested roads and traffic flow becomes smoother. In a case that the alternative path may also be congested, the iterations are repeated until the constant number of vehicles appears during the most of the entire simulation time, as shown in Figure 1.

A traffic distribution, which is a histogram of number of road segments vs. number of traffics, without and with seven iterations, of a simulation result with above settings, is shown in Figure 2. Regardless to the presence of iterations, a power-law distribution robustly appears.

3. path search in Kobe-city road network

In Figure 2, the traffic distribution of non-realistic situation without iteration without iterations is more power-law like in entire region, while the other has a cut off in the right-end region. This causes the distribution to distort from pure power law. It suggests that only path search may cause a power-law behavior without taking interactions between vehicles into account. However,

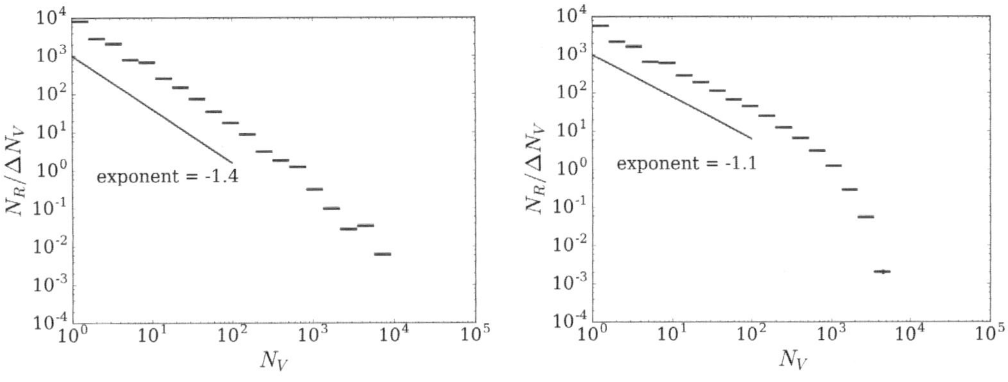

Figure 2. Traffic distribution with SUMO simulator, 1st (left) and 7th (right) iteration, quoted from [6]

this speculation requires validation, because there shall be a distortion in the distribution. As shown in Figure 1, 16,000 vehicles (74% of entire 21,600 vehicles) do not reach their destination until simulation ends due to time limit; the distribution in Figure 2 (left) is not based on either entire path or vehicles.

In purpose of path search, we used Networkx python package. We constructed directed and weighted network by extracting road connection and length from digitized map of Kobe city mentioned above. OD of vehicles are assumed to be uniform distribution, choosing each O and D from all intersections. For simplicity, we assumed uniform speed limit and no traffic signals, therefore the distance is calculated only from road lengths. Dijkstra algorithm implemented in Networkx is used for path search for each OD pair. The number of OD pairs, which correspond to the number of vehicles, is assumed to be 40,000. The entire path search of all OD pairs requires about 30 minutes to complete, using a single core of an usual PC.

We then counted how many times each path contains each road segment, and interpreted it as a number of traffics. The traffic distribution is shown in Figure 3. The distribution obeys a power law. This result evidences that a power-law behavior appears without considering interactions between vehicles at all, and it comes from network structure. Therefore, path search can be used to find out more detailed origin of the power-law behavior.

4. results in artificial road networks

In [6], we also showed the power-law behavior on randomly generated road network. Furthermore, in the previous study, we pointed out contrast of common and local roads may cause reverse-proportional relationship, which is a power-law distribution with exponent $\Gamma = -1$.

More quantitatively, we propose a toy model that a finite Cayley tree (or Bethe lattice) is embedded in the road network. An example of Cayley tree map that is used in this experiment is shown in Figure 4 (top). Assume that the network has one common road, and it branches into n_r local roads at the top of the tree. Consider n_v vehicles with uniform OD distributed in the top of the tree for simplicity. In general, local roads share n_v vehicles with their number n_r, therefore their traffic T shall be $T = n_v/n_r$ in average. The only common road is used by all these vehicles, then its traffic shall be $T = n_v$. During branching of the tree, the number of local roads n_r' shall be smaller than n_r, here, the traffic still obeys $T = n_v/n_r'$. Thus The traffic shall obey a reverse-proportional relationship $T \propto 1/n_r$, which is a power-law distribution with exponent $\Gamma = -1$. However, this speculation does not taking it into account, that there are vehicles with O or D in the way-half of the tree. Verification of this toy model with the uniform

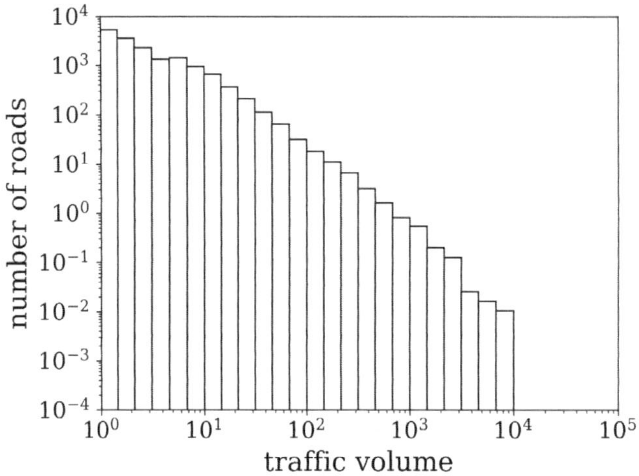

Figure 3. Traffic distribution in Kobe, with simple traffic path search, using Dijkstra method and randomized OD set, not a result of the traffic simulation.

distribution is required.

Assuming uniform distribution, it is possible to theoretically derive how many vehicles go on each edge. By removing one edge from Cayley graph can always separate the graph into completely isolated two graphs. When the smaller graph has n nodes, the other has $N - n$ nodes, where N is the number of nodes of the entire graph. The vehicles that go through this edge, the traffic T, can be derived as $T = n(N - n)$. The number of roads N_k that is connected to the nodes that have distance k from the center is given by

$$N_k = 2z(z-1)^{k-1}. \tag{1}$$

Note that it is multiplied by two because the graph is bi-directional. The traffic T_k on each of these roads is given by

$$T_k = n_k(N - n_k) = \frac{\left((z-1)^{h-k+1} - 1\right)\left(z(z-1)^h - (z-1)^{h-k+1} - 1\right)}{(z-2)^2} \tag{2}$$

because

$$n_k = \frac{(z-1)^{h-k+1} - 1}{z-2} \tag{3}$$

and

$$N = \frac{z(z-1)^h - 2}{z-2} \tag{4}$$

where n_k is the number of nodes contained in the higher part of the tree separated by the present edge, and h is the height of the tree.

We performed an exhaustive path search, by applying a complete set of OD from all nodes to all the other nodes. This is the same assumption with uniform OD set with infinite number of samples. The exhaustive traffic distribution is shown as dots in Figure 4 (bottom left). Note that there are six points, each corresponds to each layers of the tree. The theoretical curve

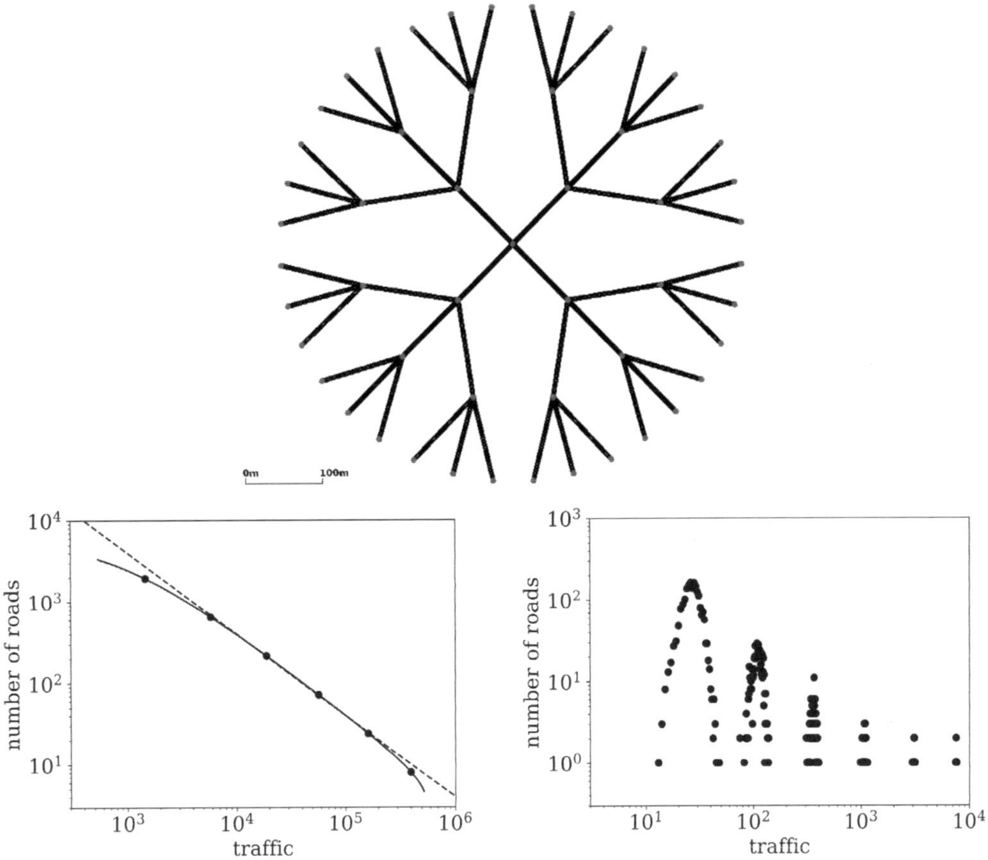

Figure 4. Top: An example of the Cayley tree with coordination number $z = 4$ and height $h = 3$. Bottom left: Points represent the theoretical discrete traffic distribution in a Cayley tree with coordination number $z = 6$, height $h = 3$, and exhaustive OD set. The solid line represents the distribution as continuum function obtained by interpreting the discrete distribution. The dashed line represents the power-law distribution with exponent -1. The theoretical distribution is power law like. Bottom right: traffic distribution in the same map as bottom left, but with a randomized OD set of 40,000 pairs.

derived in equation 1 and 2 is parametrically drawn and displayed as a solid curve for validation in Figure 4 (bottom left). A dashed line with exponent of -1 is also drawn to guide your eyes.

Also the traffic distribution with randomized OD set with 40,000 pair is shown in Figure 4 (bottom right). Note the points are diffused due to randomness. The actual road network is considered to be a combination of trees with various heights and coordination numbers. Therefore the traffic distribution shall be a superposition of these diffused distributions. It is considered to converge into a power-law like distribution.

5. summary and discussion

In the present study, we showed that simple shortest path search can cause power-law behavior in traffic distribution, using the Kobe city map and a Cayley tree. The result that path search caused the power-law behavior strongly suggests that the behavior is originated by the network structure itself, not interactions between vehicles. Also, the theoretical consideration of the power-law behavior in the Cayley tree suggests that the non-uniformity or diversity of trees that are embedded in the road network caused the power-law.

The power-law behavior robustly appears and it seems to be an aspect of an actual traffic flow. To reconstruct a realistic traffic simulation, it requires power-law appearance. The result that simple path search can cause the behavior, suggests that it is possible to construct a well-characterized urban scale traffic simulation with only such a simple algorithm, without generating detailed agents. Eventually, so much detailed information of all the agents are not required to obtain general features and responses of urban scale traffic flow. Therefore, in the future study, we are going to implement and validate a path iteration algorithm as a simplified version of iteration algorithms that are used in traffic simulators. The densities of congestions can be calculated from traffic using fundamental diagram, and the mean time required to go through each road segments, shall allow to reroute the path.

acknowledgements

We thank T. Uchitane and T. Aoki for helpful discussions. We are also grateful to development team of OACIS [7], which we used to execute and combine many simulation results. This research was supported by MEXT as Exploratory Challenges on Post-K computer (Studies of multi-level spatiotemporal simulation of socioeconomic phenomena).

References

[1] Schreckenberg, M., Schadschneider, A., Nagel, K., & Ito, N. (1995). Discrete stochastic models for traffic flow. *Physical Review E, 51*(4), 2939.
[2] Uchitane, T., & Ito, N. (2016). Applying Factor Analysis to Describe Urban Scale Vehicle Traffic Simulation Results. *Transactions of the Society of Instrument and Control Engineers*, 52, 545-554. (in Japanese)
[3] Asano, Y., Ito, N., Inaoka, H., Imai, T., & Uchitane, T. (2015). Traffic Simulation of Kobe-City. *In Proceedings of the International Conference on Social Modeling and Simulation, plus Econophysics Colloquium 2014* (pp. 255-264). Springer, Cham.
[4] Okuyama, K., Takayasu, M., & Takayasu, H. (1999). Zipf's law in income distribution of companies. *Physica A: Statistical Mechanics and its Applications*, 269(1), 125-131.
[5] Adamic, L., & Huberman, B. (1999). *The nature of markets in the World Wide Web.*
[6] Umemoto, D., & Ito, N. (2018). Power-law distribution in an urban traffic flow simulation. *Journal of Computational Social Science*, 1(2), 493-500.
[7] Murase, Y., Uchitane, T., & Ito, N. (2014). A tool for parameter-space explorations. *Physics Procedia*, 57, 73-76.
[8] Kesten, H. (1982). Percolation theory for mathematicians (Vol. 423). *Boston: Birkhuser.*

Comparison of quantum and classical methods for labels and patterns in Restricted Boltzmann Machines

Vivek Dixit[1], **Yaroslav Koshka**[1,2], **Tamer Aldwairi**[3], **M.A. Novotny**[1,4]

[1]CCS, HPCC, Mississippi State University (MSU), Mississippi State, MS 39759, USA
[2]Department of Electrical and Computer Engineering, MSU, Mississippi State, MS 39762, USA
[3]Mathematics and Computer Science, Ursinus College, Collegeville, PA 19426, USA
[4]Department of Physics and Astronomy, MSU, Mississippi State, MS 39762, USA

E-mail: man40@msstate.edu

Abstract. Classification and data reconstruction using a restricted Boltzmann machine (RBM) is presented. RBM is an energy-based model which assigns low energy values to the configurations of interest. It is a generative model, once trained it can be used to produce samples from the target distribution. The D-Wave 2000Q is a quantum computer which has been used to exploit its quantum effect for machine learning. Bars-and-stripes (BAS) and cybersecurity (ISCX) datasets were used to train RBMs. The weights and biases of trained RBMs were used to map onto the D-Wave. Classification as well as image reconstruction were performed. Classification accuracy of both datasets indicates comparable performance using D-Wave's adiabatic annealing and classical Gibb's sampling.

1. Introduction

D-Wave 2000Q is D-Wave's most advanced quantum machine, and is well suited to discrete optimization problems. It finds solution to a problem by finding ground state of the corresponding classical Ising spin model. For a user defined classical Ising spin glass, it returns a state using quantum adiabatic procedure. The returned state is hopefully a ground state. It is a new tool, and researchers are trying to figure out applications by comparing its performance with other existing approaches and techniques.

Mott *et al* [1] used D-Wave to classify Higgs-boson-decay signals vs. background. They showed that the quantum and classical annealing-based classifiers perform comparable to the state-of-the-art machine learning methods. Mniszewski *et al* [2] found that the results for graph partitioning using D-Wave systems were comparable to current state-of-the-art methods. Alexandrov *et al* [3] used a D-Wave for matrix factorization. Kais *et al* [4] showed an exact mapping between the electronic structure Hamiltonian and the Ising Hamiltonian. Lidar *et al* [5] used a D-Wave for classification of DNA sequences according to their binding affinities. Koshka *et al* [6, 7, 8, 9] performed classification and image reconstruction of a bars and stripes dataset and explored the energy landscape using a RBM embedded onto the D-Wave.

The objective of this work is to investigate D-Wave's QPU (Quantum Processing Unit) performance on bars-and-stripes (BAS) and cybersecurity (ISCX) datasets. The BAS dataset was chosen for its simplicity, the correct ground states can be visually identified. Classification and reconstruction on BAS dataset demonstrated in our previous work [6], has been reproduced here in order to investigate how the same approach works for more complex ISCX dataset, which should be expected to result in a more complicated energy landscape. The ISCX dataset was generated by the Canadian Institute for Cybersecurity. It has been recently shown that the attack and benign records of the ISCX dataset can be classified using a RBM [10]. In this work, we compare the classification performance of a regular RBM which uses Gibb's sampling and a RBM embedded on the D-Wave.

2. Methods

2.1. Restricted Boltzmann Machine (RBM)

A RBM is a generative artificial neural network that can learn a probability distribution over multiple variables. It consists of a layer of observable variables (visible layer) and a layer of latent variables (hidden layer). Connections between the variables in the same layer are not permitted. Let v and h be vectors of variables in the visible and hidden layer, respectively. The energy of the model is given as:

$$E(v,h) = -b^T v - c^T h - v^T W h, \tag{1}$$

where 'b' and 'c' are bias vectors at the visible and hidden layers, respectively. 'W' is the weight matrix of the model parameters. A RBM is an energy based model, with the probability distribution, $P(v,h)$, a function of the energy of the configuration,

$$P(v,h) = \frac{1}{Z}e^{-E(v,h)}, \qquad Z = \sum_v \sum_h e^{-E(v,h)}. \tag{2}$$

The sum over all possible v and h vectors makes the partition function Z very difficult to evaluate. The joint probability, $P(v,h)$, being a function of Z is also hard. Due to the bipartite graph structure of the RBM, the conditional distributions $P(h|v)$ and $P(v|h)$ are simple to compute,

$$P(h|v) = \prod_{i=1}^{m} P(h_i|v), \qquad P(v|h) = \prod_{j=1}^{n} P(v_j|h). \tag{3}$$

where 'm' and 'n' are the number of units in the hidden and visible layers. The individual activation probabilities are given by:

$$P(h_j = 1|v) = \sigma\Big(c_j + \sum_{i=1}^{n} w_{i,j} v_i\Big), \qquad P(v_i = 1|h) = \sigma\Big(b_i + \sum_{j=1}^{m} w_{i,j} h_i\Big) \tag{4}$$

where σ is the sigmoid function. A RBM is trained by maximizing the probability distribution over the training data. The log likehood is given by:

$$l(W,b,c) = \sum_{t=1}^{n} logP\left(v^{(t)}\right) = \sum_{t=1}^{n} log \sum_{h} P\left(v^{(t)}, h\right) \tag{5}$$

where $v^{(t)}$ is a sample in the training dataset and

$$l(W,b,c) = \sum_{t=1}^{n} log \sum_{h} e^{-E(v^{(t)},h)} - n \cdot log \sum_{v,h} e^{-E(v,h)}. \tag{6}$$

Denote $\theta = \{W, b, c\}$. The gradient of the log-likelihood is given by:

$$\nabla_\theta l(\theta) = \sum_{t=1}^{n} \frac{\sum_h e^{-E(v^{(t)},h)} \nabla_\theta(-E(v^{(t)},h))}{\sum_h e^{-E(v^{(t)},h)}} - n \cdot \frac{\sum_{v,h} e^{-E(v,h)} \nabla_\theta(-E(v,h))}{\sum_{v,h} e^{-E(v,h)}} \tag{7}$$

$$\nabla_\theta l(\theta) = \sum_{t=1}^{n} \mathbb{E}_{P(h|v^{(t)})}[\nabla_\theta(-E(v^{(t)},h))] - n \cdot \mathbb{E}_{P(h|v)}[\nabla_\theta(-E(v,h))]. \tag{8}$$

Once we have an equation for the gradient of the log likelihood, weights and biases can be estimated using gradient accent optimization: $\theta_j^{new} = \theta_j^{old} + \epsilon \cdot \nabla_{\theta_j} l(\theta_j)$, where ϵ is the learning rate.

2.2. D-Wave System

The Hamiltonian for a D-Wave system of qubits can be represented as:

$$H_{Ising} = -\frac{A(s)}{2}\Big(\sum_i \hat{\sigma}_x^i\Big) + \frac{B(s)}{2}\Big(\sum_i h_i \hat{\sigma}_z^i + \sum_{(i>j)} J_{ij} \hat{\sigma}_z^i \hat{\sigma}_z^j\Big) \tag{9}$$

where $\sigma_{x,z}^{(i)}$ are Pauli matrices operating on qubit q_i, and h_i and $J_{i,j}$ are the qubit biases and coupling strengths. s is called the anneal fraction. $A(s)$ and $B(s)$ are known as anneal functions. At $s = 0$, $A(s) \gg B(s)$, while $A(s) \ll B(s)$ for $s = 1$. As we increase s from 0 to 1, anneal functions change gradually to meet these boundary conditions. In the standard quantum annealing protocol, s changes from 0 to 1. The network of qubits starts in a global superposition over all possible classical states and after $s = 1$, the system is measured in a single classical state.

The arrangement of qubits on the D-Wave chip forms a C16 Chimera graph: 2048 qubits are mapped into a 16×16 matrix of unit cells each of 8 qubits. Figure 1(a) shows a C3 Chimera graph with 3×3 unit cells. Within each unit cell there are two set of 4 qubits which are connected to each other in a bipartite fashion. Each qubit in a unit cell connects to two qubits of the other unit cells. Thus, each qubit connects to a maximum of 6 other qubits. This connectivity can be enhanced by forming ferromagnetic bonds between the qubits as shown in the Figure 1. We follow the embedding approach that was successfully utilized in our previous work [6, 7, 8, 9]. One visible (hidden) unit of a RBM is formed by connecting 16 vertical (horizontal) qubits. There are 64 sets of vertical and horizontal qubits on a C16 Chimera graph. Thus, we get a RBM of 64 visible and hidden units by forming chains in this fashion. Of course care must be taken for any inaccessible qubits to form a further restriced RBM.

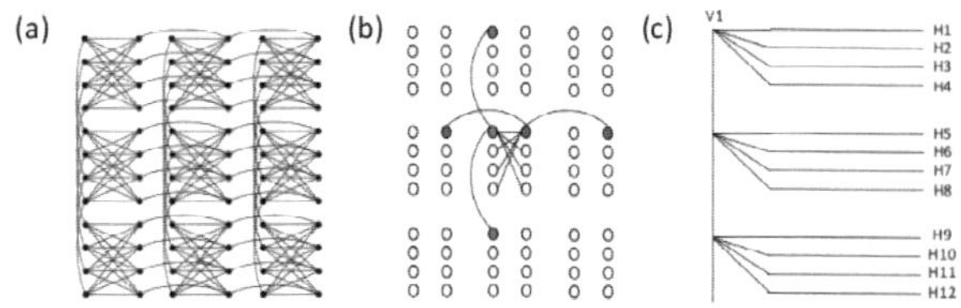

Figure 1. (a) C3 Chimera graph of qubits. (b) Several vertical (red) and horizontal (blue) qubits are chained to form a visible and a hidden unit. (c) Connectivity of visble unit (V1) with 12 hidden units. Each unit is formed by ferromagnetic bonds between three qubits.

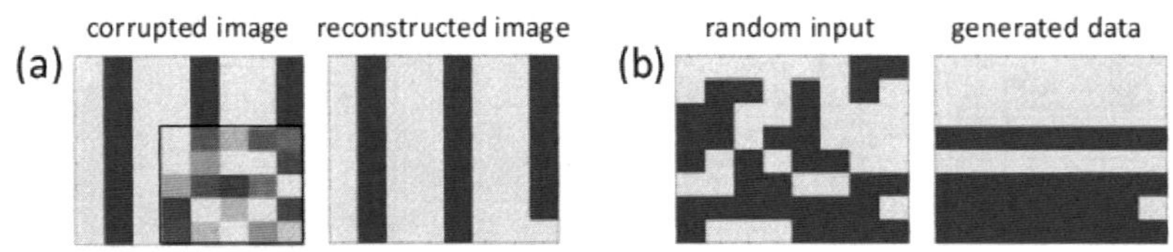

Figure 2. (a) Input and output of a trained RBM used for image reconstruction. (b) RBM used as a generative model. Input is random noise and output is a labeled stripe pattern.

3. Results and Discussion

Two datasets (BAS and ISCX) were used for the RBM training and successively embedding onto the D-Wave. BAS is a binary dataset with each record of 64 bit length. The last two bits represent the label of the record: 01 for a bar pattern and 10 for a stripe (Figure 4). Our BAS dataset was comprised of 500 unique records. The number of unique samples used for training was 300, with the remaining 200 samples held for testing the classification accuracy. The ISCX dataset was generated from a bigger dataset obtained from the 'Canadian Institute for Cybersecurity' using feature selection [10]. It has 9800 unique records, each record was design to be of length 64 bits. The last two bits represent the label of the record: normal network traffic (01) and malicious network traffic or attacks (10). 7000 records were used for training and the remaining records for the testing.

First, a RBM was trained on the BAS dataset. Weights and biases of the trained RBM were used to embed it onto the D-Wave chip. 95.5% of bars and 94.5% stripes patterns are correctly predicted. Image reconstruction and data generation were also performed using the trained RBM. For image reconstruction, first 25 bits of a record was corrupted, then fed to the RBM framework. After a few Gibb's cycles, the output was sampled from the visible units. It was found that in almost all the cases full image reconstruction was achieved. The corrupted and reconstructed images are shown in figure 2 (a). For data generation, a random input vector of 0s and 1s was fed to the RBM. The output was a pattern similar to a pattern in the BAS dataset. An example of random input and sampled output is shown in Figure 2(b). Next, the cybersecurity, ISCX, dataset was used to train a RBM. ISCX test dataset had 1401 benign and 1399 attack records. 84.4% of benign and 83.3% of attack records were correctly predicted.

Weights and biases obtained from the trained models were then used for D-Wave embedding. Weights were used as coupling strengths $J_{i,j}$ and biases specified the local fields h_i of the qubits, thus forming the objective function. As described earlier, qubits were combined to form RBM units. The embedded RBM had 64 visible and 64 hidden units. We refer to this embedded RBM as a quantum RBM, while the regular RBM as a classical RBM. Once the embedding was

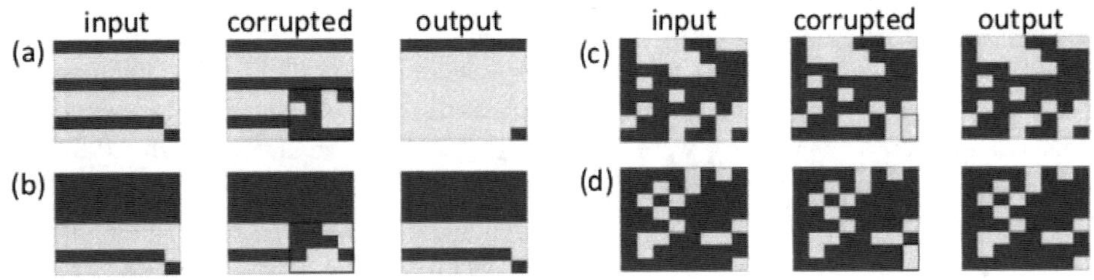

Figure 3. (a,b) BAS dataset image reconstruction, and (c,d) classifying label reconstruction of ISCX samples using quantum RBM. 64 bits ISCX records are shown as 8×8 images.

complete, a bias vector h was applied to the visible units of the quantum RBM. The components of this bias vector were proportional to the input sample data. At this step noise was introduced to the bias vector in order to hide the classifying labels or to corrupt the image. A forward anneal was performed and the final state of the qubits were obtained. The final state of the corrupted unit gave the classification/reconstruction result. This procedure of image/data reconstruction is the same as used in the earlier work [6]. In the case of the BAS dataset image reconstruction was performed using the quantum RBM. In most cases reconstruction was easily performed and the output was identical to the uncorrupted image, however, in a few cases a similar but non-identical pattern was obtained as shown in Figure 3(a). Classification was performed on the cybersecurity (ISCX) dataset. Figure 3(c) and 3(d) show two examples where classification of the ISCX samples were performed using the quantum RBM.

To compare performance of the classical RBM and quantum RBM (or D-Wave embedded RBM), we calculated the percentage classification accuracy on both the BAS and ISCX datasets. Accuracy is given by:

$$Accuracy = \frac{TP + TN}{TP + FP + TN + FN} \times 100.$$

T and F stand for true and false; P and N stand for positive and negative. Results are shown in Table 1. Based on these results, we conclude that classification performances of the quantum and classical RBMs are comparable. In other words, minimization of the objective function using adiabatic quantum annealing gave the similar results as Gibb's sampling of the classical RBM for both BAS and ISCX datasets. This also indicates that the D-Wave optimizer works as well as classical methods for these datasets.

Table 1. Percentage classification accuracy of classical RBM (Gibb's sampling) and quantum RBM (quantum annealing).

Dataset	Accuracy	
	classical RBM	quantum RBM
BAS	94.0%	94.5%
ISCX	83.9%	82.2%

4. Conclusions

RBMs were trained on bar-and-stripes (BAS) and cybersecurity datasets using contrastive divergence (CD-1). Classification accuracy was measured for the both datasets. Image reconstruction and data generation were performed for the BAS dataset. In addition to the previously demonstrated ability to successfully embed RBM trained on a toy example of BAS dataset, this work shows an effective embedding of a RBM trained on a much more complicated cybersecurity dataset. Image reconstruction was performed and classification accuracy were estimated for the embedded RBM. In the final recognition step, the quantum RBM uses the quantum adiabatic annealing, while Gibb's sampling is used in the classical RBM. Both the classical and quantum RBM showed comparable classification accuracy. This result is in-line with observations of other researchers [1, 2, 3, 4, 5] where the D-Wave performed in a manner comparable to the existing state-of-the-art methods.

Acknowledgments

The authors would like to acknowledge D-Wave Systems for their use of the Burnaby D-Wave machine. This material is based on research sponsored by the Air Force Research Laboratory (AFRL) under agreement number FA8750-18-1-0096. The U.S. Government is authorized to reproduce and distribute reprints for Governmental purposes notwithstanding any copyright notation thereon. The views and conclusions contained herein are those of the authors and should not be interpreted as necessarily representing the official policies or endorsements, either expressed or implied, of the Air Force Research Laboratory (AFRL) or the U.S. Government.

References

[1] Mott A, Job J, Vlimant J R, Lidar D and Spiropulu M 2017 *Nature* **550** 375–379 URL `https://doi.org/10.1038/nature24047`

[2] Ushijima-Mwesigwa H, Negre C F A and Mniszewski S M 2017 *Proceedings of the Second International Workshop on Post Moores Era Supercomputing* PMES'17 (New York, NY, USA: ACM) pp 22–29 ISBN 978-1-4503-5126-3 URL `http://doi.acm.org/10.1145/3149526.3149531`

[3] OMalley D, Vesselinov V V, Alexandrov B S and Alexandrov L B 2018 *PLOS ONE* **13** 1–12 URL `https://doi.org/10.1371/journal.pone.0206653`

[4] Xia R, Bian T and Kais S 2018 *The Journal of Physical Chemistry B* **122** 3384–3395 pMID: 29099600 (*Preprint* `https://doi.org/10.1021/acs.jpcb.7b10371`) URL `https://doi.org/10.1021/acs.jpcb.7b10371`

[5] YLi R, Felice R D, Rohs R and ALidar D 2018 *npj Quantum Information* **4** URL `https://doi.org/10.1038/s41534-018-0060-8`

[6] Koshka Y and Novotny M A 2018 *IEEE Transactions on Emerging Topics in Computational Intelligence* 1–11 ISSN 2471-285X

[7] Koshka Y and Novotny M A 2018 *2018 International Joint Conference on Neural Networks (IJCNN)* pp 1–8 ISSN 2161-4407

[8] Koshka Y, Perera D, Hall S and Novotny M A 2017 *Neural Computation* **29** 1815–1837 pMID: 28562219

[9] Koshka Y, Perera D, Hall S and Novotny M A 2016 *2016 International Joint Conference on Neural Networks (IJCNN)* pp 1948–1954 ISSN 2161-4407

[10] Aldwairi T, Perera D and Novotny M A 2018 *Computer Networks* **144** 111 – 119 ISSN 1389-1286 URL `http://www.sciencedirect.com/science/article/pii/S1389128618306005`

Off-lattice and parallel implementations of the pivot algorithm

Nathan Clisby and Dac Thanh Chuong Ho

Department of Mathematics, Swinburne University of Technology, P.O. Box 218, Hawthorn, Victoria 3122, Australia

E-mail: nclisby@swin.edu.au

Abstract. The pivot algorithm is the most efficient known method for sampling polymer configurations for self-avoiding walks and related models. Here we introduce two recent improvements to an efficient binary tree implementation of the pivot algorithm: an extension to an off-lattice model, and a parallel implementation.

1. Introduction

Self-avoiding walks are non-intersecting paths on lattices such as the two-dimensional square lattice or the three-dimensional simple cubic lattice. Due to universality, they exactly capture the essential physics of the excluded-volume effect for polymers in the good-solvent limit, and as such can be used to study features such as the value of the Flory exponent ν which relates the geometric size of a walk to the number of monomers in the chain.

The pivot algorithm is the most efficient known method for sampling self-avoiding walks of fixed length. It is a Markov chain Monte Carlo method, which was invented by Lal [1], but first studied in depth by Madras and Sokal [2], who also invented an efficient hash table implementation. Recent improvements to the implementation of the pivot algorithm [3–5] have dramatically improved computational efficiency to the point where it is possible to rapidly sample polymer configurations with up to 1 billion monomers [6].

In this paper, we will describe two recent improvements in algorithms to sample self-avoiding walks, focusing in particular on the pivot algorithm. In Sec. 2 we describe an off-lattice implementation of the SAW-tree data structure [5]. In Sec. 3 we describe a parallel implementation of the pivot algorithm which improves the sampling rate for very long walks. Finally, we have a brief discussion about prospects for further progress and conclude in Sec. 4.

2. Off-lattice implementation

The SAW-tree data structure [5] is a binary tree that encodes information about the self-avoiding in an efficient way in nodes in the tree. In particular, the leaves of the tree consist of individual monomers, while the internal nodes store aggregate information about all of the monomers that are below that node within the tree, as well as "symmetry" information which encodes transformations that must be applied to sub-walks before they are concatenated together. The aggregate information that must be stored includes information about the extent of the sub-walk in the form of a "bounding volume", which is taken to be a rectangle for square-lattice walks, and a rectangular prism for simple-cubic-lattice walks. For lattice self-avoiding walks, the symmetry elements are rotations and reflections that leave the lattice invariant. See [5] for a full description of the implementation.

Although lattice self-avoiding walks capture the universal behaviour of polymers in the good-solvent limit, there are strong arguments for why off-lattice models of polymers may have advantages under certain circumstances. Firstly, they provide an opportunity to empirically model more realistic interactions for polymers, and thus to reproduce not only universal features but also make precise experimental predictions. Secondly, under some circumstances it may be the case that the effect of the lattice may have a non-negligible effect, for example when trying to understand the nature of the globule transition it may be the case that the restriction to the lattice significantly influences the nature of the transition. Finally, while lattices have discrete symmetry groups, the symmetry group corresponding to reflections and rotations of $\mathbb{R}^d$ is the continuous orthogonal group $O(d)$. This continuous group allows for more freedom for performing pivot moves, and it is conceivable that this additional freedom may enhance sampling efficiency under some circumstances.

We implement the SAW-tree for the bead-necklace, or tangent-hard-sphere, model, which consists of a fully flexible chain of hard spheres that just touch. A typical configuration for this model in $\mathbb{R}^2$ is shown in Fig. 1.

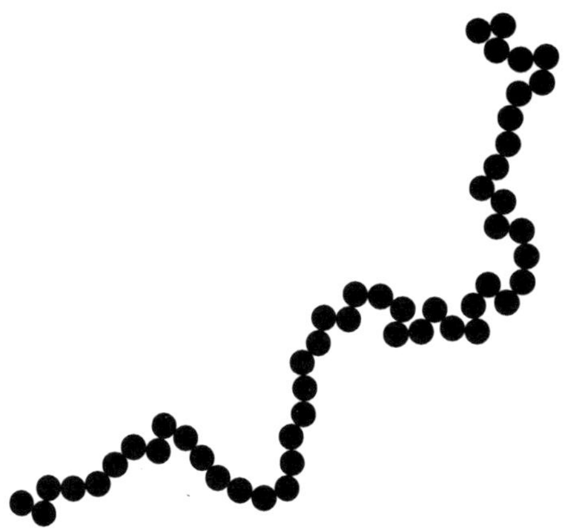

Figure 1. Typical bead-necklace configuration with 50 monomers in $\mathbb{R}^2$.

We will now describe the key features of our implementation, and will present evidence that the off-lattice SAW-tree implementation of the pivot algorithm has $O(\log N)$ performance in line with the performance of the original lattice SAW-tree implementation. The description will not be self-contained, and the interested reader is referred to [5] for relevant details.

The orthogonal group $O(2)$ is used as the symmetry group for $\mathbb{R}^2$, and similarly $O(3)$ is used for $\mathbb{R}^3$. The orthogonal group includes rotations as the subgroups $SO(2)$ and $SO(3)$ respectively, but also includes reflection moves.

Symmetry group elements are sampled uniformly at random so as to preserve the Haar measure [7] on the group. This automatically ensures that the Markov chain satisfies the detailed-balance condition, and so must be sampling configurations with uniform weights.

As for ergodicity, we feel that it is extremely likely that the algorithm is ergodic. For lattice models the pivot algorithm has been proved to be ergodic; this was first done for $\mathbb{Z}^2$ and $\mathbb{Z}^3$ in the seminal paper of Madras and Sokal [2]. Interestingly, inclusion of reflections seem to be necessary for ergodicity for lattice models. In the continuum, it is our view that the additional

Journal of Physics: Conference Series **2122** (2021) 012008 doi:10.1088/1742-6596/2122/1/012008

freedom afforded as compared to the lattice should mean that pivot algorithm is ergodic in this case, too. We do not have sufficient insight into the problem to know whether the extra freedom would allow one to have an ergodic algorithm with only rotations (and not reflections). Some theoretical work has been done previously on the ergodicity of pivot moves for continuous models [8], but this is not directly relevant here as the proof relied on double-pivot moves.

The key decision for the SAW-tree implementation for the bead-necklace model is the choice of *bounding volume* to be used. The bounding volume is a shape which is stored in nodes in the SAW-tree, such that it is guaranteed that the entire sub-chain which is represented by the node is completely contained within the bounding volume. The use of a bounding volume is necessary for the rapid detection of self-intersections when a pivot move is attempted.

The natural choice of the bounding volume for $\mathbb{Z}^2$ is the rectangle, and for $\mathbb{Z}^3$ the natural choice is the rectangular prism. This is because these shapes snugly fit the sub-chains that they contain (in the sense that the sub-chains must touch each boundary or face of the shape), and the shapes are preserved under lattice symmetry operations.

The natural shape for the bounding volume for the bead-necklace model for $\mathbb{R}^2$ would seem to be the circle, and similarly for $\mathbb{R}^3$ the natural choice would be the sphere. This is because these are the only shapes that are invariant under the action of $O(2)$ and $O(3)$ respectively.

One of the operations that must be performed with bounding volumes [5] is the merge operation, which involves combining two bounding volumes (which contain sub-chains) to create a bounding volume that contains both of the original bounding volumes (and hence contains both sub-chains). In contrast to the situation for lattice models, the bounding volumes which result from the merge operation do not necessarily form a snug fit for the polymer sub-chains. This is illustrated in Fig. 2 for an example in $\mathbb{R}^2$ where the snugly fitting bounding circles for two sub-chains are merged together so that they contain the concatenated walk. The concatenated walk *does not* touch the boundary of the larger circle.

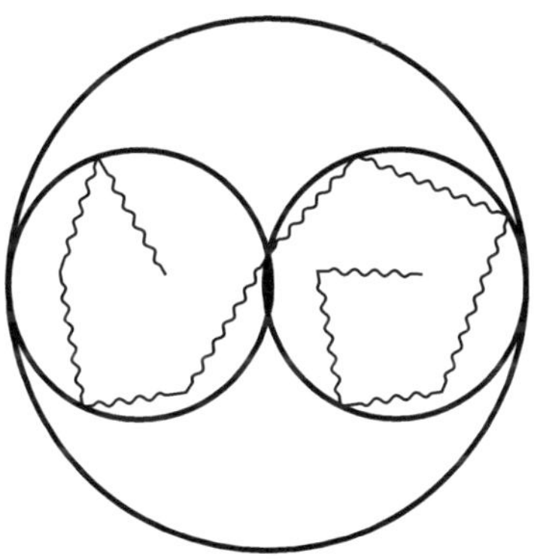

Figure 2. Merging bounding circles for two sub-chains to form the bounding circle for the concatenated chain in $\mathbb{R}^2$.

A priori, we had no expectation about whether the lack of snug fit for the bounding volumes would prove to be a significant problem. We considered it possible that the error from the fit would grow rapidly as one moved up the SAW-tree, and this would have worsened the performance of the intersection testing algorithm. But, we found that in fact this was not a problem at all. We estimated the mean ratio of the diameter of the bounding volume to the square root of the mean value of the squared end-to-end distance $\langle R_E^2 \rangle^{1/2}$. We found that as the length of the chains increased the ratio was approaching a constant for both $\mathbb{R}^2$ and $\mathbb{R}^3$, indicating that the error was becoming saturated. For chain lengths of $N = 10^6$ this ratio was only 1.45 for $\mathbb{R}^2$, and 1.71 for $\mathbb{R}^3$. Thus, in the average case this suggests that the lack of a snug fit only results in a constant factor error in the diameter of the bounding volume for the off-lattice implementation. This means that the behaviour of the lattice and off-lattice implementations should be essentially the same, up to a constant factor.

We evaluated the mean CPU time per pivot move for a range of polymer lengths, for lattice and off-lattice SAW-tree implementations in two and three dimensions on Dell PowerEdge FC630 machines with Intel Xeon E5-2680 CPUs, and plot the results of these computer experiments in Figs 3 and 4.

We found that the time per pivot move attempt was somewhat worse for the off-lattice implementation as compared to the lattice implementation, which was to be expected due to the increased number of operations required for computations involving the symmetry elements and coordinate vectors. But, in absolute terms the performance is still impressive, and for polymers with 10^7 monomers pivot attempts are performed in mean CPU time of less than 6μs for $\mathbb{R}^2$, and in less than 40μs for $\mathbb{R}^3$. We clearly observe $O(\log N)$ behaviour in each case, which is strong evidence that the off-lattice implementation behaves in fundamentally the same way as the original lattice implementation of the SAW-tree.

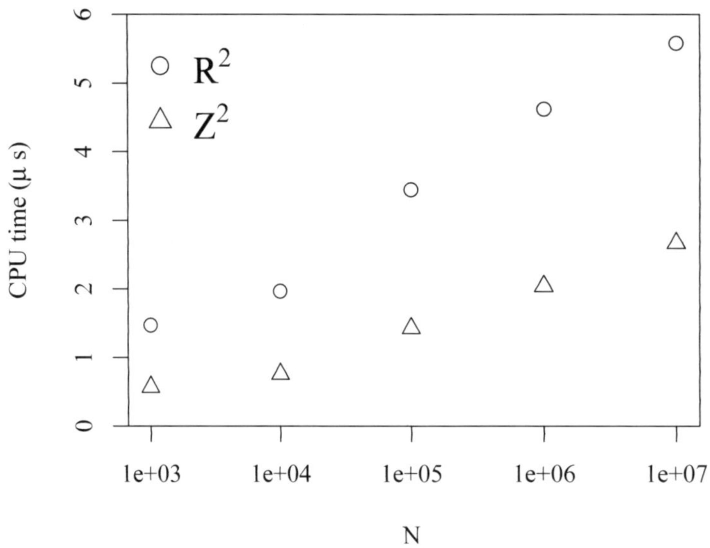

Figure 3. CPU time per pivot move attempt for the bead-necklace model in $\mathbb{R}^2$, in comparison to SAWs in $\mathbb{Z}^2$, plotted against the number of monomers N.

3. Parallel implementation of the pivot algorithm

The SAW-tree implementation of the pivot algorithm [5] is remarkably efficient, but it suffers from one significant drawback: the intersection testing and SAW-tree update procedures are inherently serial operations. This makes it difficult to take advantage of additional cores to improve the rate at which polymer configurations are sampled. To some extent this issue is obviated by the fact that for number of monomers N up to the order of tens of millions or even 100 million it is possible to run simulations in parallel on multicore machines, and still obtain results in a reasonable clock time.

But, in the regime where a large amount of memory is needed for truly large N, of the order of $10^8 - 10^9$, on the Dell PowerEdge FC630 machines with Intel Xeon E5-2680 CPUs on which

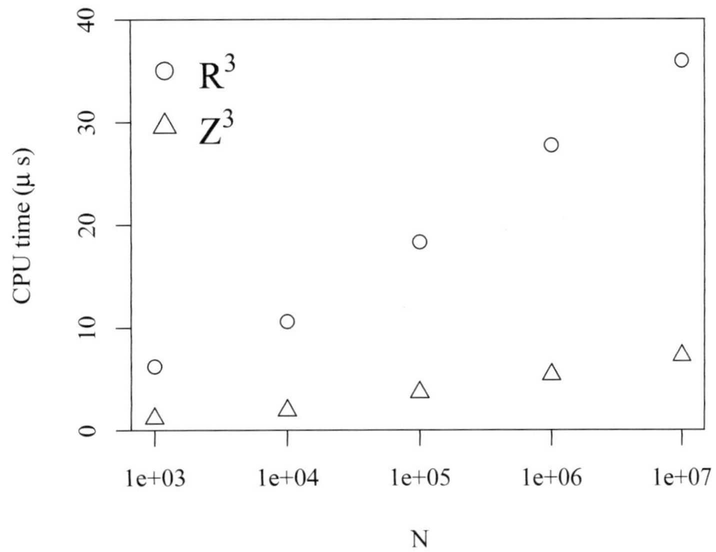

Figure 4. CPU time per pivot move attempt for the bead-necklace model in $\mathbb{R}^3$, in comparison to SAWs in $\mathbb{Z}^3$, plotted against the number of monomers N.

computer experiments are being run this prevents all cores being simultaneously used due to memory constraints[1]. Under these circumstances most cores must be left idle while data is being collected.

Here we will briefly sketch a method to improve the sampling rate by utilising additional cores in exactly this difficult regime.

The key insight is that as the number of monomers increases, the probability of a pivot move being successful decays as a power law of the form N^{-p}, with $p \approx 0.19$ for $\mathbb{Z}^2$, and $p \approx 0.11$ for $\mathbb{Z}^3$. For $N = 10^9$ on $\mathbb{Z}^2$, the probability of a pivot move being successful is 0.019, which means that on average roughly 50 unsuccessful pivot attempts are made for each success.

Given that most proposed pivot moves in this regime fail, and so do not result in any update being made for the self-avoiding walk, it is possible to perform many pivot attempts in parallel without this effort being wasted.

For example, imagine that we are sampling SAWs of 10^9 steps via the pivot algorithm, and we may test for success or failure of up to ten pivot moves simultaneously. Note that a move consists of a proposed monomer location to act as the centre of the pivot move, and a proposed symmetry operation. Suppose for the first batch of ten proposed moves $\{M_1, M_2, \cdots, M_{10}\}$, that each of these moves were unsuccessful. Then, we can move on to another batch, and none of the work performed by any of the threads was wasted. Suppose for the second batch $\{M_{11}, M_{12}, \cdots, M_{20}\}$ that the first 6 moves $M_{11}, \cdots, M_{16}$ are unsuccessful, but M_{17} is successful. Then we need to perform the update associated with the move M_{17} which must happen as a serial operation performed by a single thread. It does not matter whether M_{18}, M_{19}, M_{20} were successful or not: these tests will need to be performed again in case the update has altered the result of the test. The next batch will then consist of ten proposed moves $\{M_{18}, M_{19}, \cdots, M_{27}\}$.

The tests for success or failure will occur for each thread regardless of the outcome of the tests performed by other threads. But, provided the probability of multiple successful moves occurring in a batch is low, then most of this work will not be wasted. The lower the probability

[1] There are 24 cores, and total memory available is 128GB.

of success, the greater the potential for speed up to occur by exploiting parallelism.

We have implemented this idea in a prototype C program with OpenMP being used for managing the parallel pivot attempts. The SAW-tree is held in shared memory where all threads can access it for performing intersection tests. When a pivot move is found to be successful, then the update is performed by a single thread while all other threads remain idle.

We performed computer experiments to test this implementation on the aforementioned FC630 machines for SAWs of various lengths on the square lattice. We utilised 24 threads, with batches (or chunks) of 48 pivot attempts which meant that each thread made two attempted pivot moves on average. We collated the calendar time per pivot attempt in μs in Table 1. The value t_1 is the mean CPU time for a single thread, while t_{24} is the mean CPU time for the 24 threads running in parallel. We see that as N increases the probability of a move being successful decreases, and the relative performance of the parallel implementation to the serial implementation improves. For $N = 10^9$ there is roughly a four-fold improvement in performance.

Although it is suitable as a proof-of-concept, the implementation developed thus far is only a prototype, and more work remains to be done to improve its performance. In particular, it should be possible to re-use some information from intersection tests even if these moves are scheduled to occur after a move that is found to be successful. For example, if a move is found to cause a self-intersection between monomers labelled l and m along the chain, then if the prior succesful move involved a pivot site outside of the interval l to m then this would not have any effect on the self-intersection. Nonetheless, even in its current state the performance gain is sufficient to make it worthwhile for use in the large N, memory-limited regime.

Table 1. Performance of the prototype parallel implementation of the pivot algorithm for $\mathbb{Z}^2$.

N	Pr(success)	1/ Pr(success)	t_1 (μs)	t_{24} (μs)	t_1/t_{24}
10^6	0.068	15	1.58	1.35	1.17
10^7	0.044	23	2.31	1.07	2.16
10^8	0.029	34	2.90	0.903	3.21
10^9	0.019	53	3.16	0.805	3.93

4. Discussion and conclusion

Schnabel and Janke [9] have very recently implemented a binary tree data structure which is similar to the SAW-tree for the bead-necklace model, as well as a model for which the Lennard-Jones interaction is implemented. The implementation for the bead-necklace model appears to have roughly the same computational efficiency as the implementation sketched here. The efficient implementation for the Lennard-Jones polymer model is very interesting, and a significant advance on the state of the art. It will be interesting to see if further progress in this direction can be made, for example in the evaluation of Coulomb interactions which would be necessary for efficient simulation of polyelectrolytes.

Full details for the off-lattice SAW-tree implementation of the pivot algorithm will be presented elsewhere in future.

More work needs to be done to test and improve the implementation of the parallel version of the pivot algorithm. In future, the parallel implementation of the pivot algorithm will allow for improved simulations of very long SAWs on the square lattice. The method will result in significant speed-ups for SAWs with hundreds of millions or even one billion steps, especially for the square lattice.

References

[1] Lal M 1969 *Mol. Phys.* **17** 57–64
[2] Madras N and Sokal A D 1988 *J. Stat. Phys.* **50** 109–186
[3] Kennedy T 2002 *J. Stat. Phys.* **106** 407–429
[4] Clisby N 2010 *Phys. Rev. Lett.* **104** 055702
[5] Clisby N 2010 *J. Stat. Phys.* **140** 349–392
[6] Clisby N 2018 *J. Stat. Phys.* **172** 477–492
[7] Stewart G W 1980 *SIAM Journal on Numerical Analysis* **17** 403–409
[8] Plunkett L and Chapman K 2016 *J. Phys. A: Math. Theor.* **49** 135203
[9] Schnabel S and Janke W (*Preprint* `arXiv:1904.11191`)

Acknowledgements

Thanks to Stefan Schnabel for communicating results regarding an alternative efficient off-lattice implementation of the pivot algorithm prior to publication. N.C. gratefully acknowledges support from the Australian Research Council under the Future Fellowship scheme (project number FT130100972).

Universal finite-size scaling function for coarsening in the Potts model with conserved dynamics

Wolfhard Janke[1], **Suman Majumder**[1] **and Subir K. Das**[2]

[1]Institut für Theoretische Physik, Universität Leipzig, Postfach 100 920, 04009 Leipzig, Germany

[2]Theoretical Sciences Unit and School of Advanced Materials, Jawaharlal Nehru Centre for Advanced Scientific Research, Jakkur P.O., Bangalore 560064, India

E-mail: `janke@itp.uni-leipzig.de,majumder@itp.uni-leipzig.de,das@jncasr.ac.in`

Abstract. We study kinetics of phase segregation in multicomponent mixtures via Monte Carlo simulations of the q-state Potts model, in two spatial dimensions, for $2 \le q \le 20$. The associated growth of domains in finite boxes, irrespective of q and temperature, can be described by a single universal finite-size scaling function, with only the introduction of a nonuniversal metric factor in the scaling variable. Our results show that although the scaling function is independent of the type of transition, the q-dependence of the metric factor hints to a crossover at $q = 5$ where the type of transition in the model changes from second to first order.

1. Introduction

The nonequilibrium process of phase separation of a mixture into its components, following a quench inside its demixing region, happens through the growth of clusters or domains of like species, once the system becomes thermodynamically unstable [1, 2]. The characteristic dynamic length scale of the process, i.e., the average domain size ℓ, generally follows a power law with time t, viz., $\ell(t) \sim t^{\alpha}$, α being the growth exponent. Typically, α depends on the intrinsic dynamics of the system, e.g., in solid mixtures, where the dynamics is diffusive, the value of α is $1/3$, referred to as the Lifshitz-Slyozov (LS) growth law [3]. To understand segregation phenomena in two-component solids, the spin-1/2 Ising Hamiltonian has been extensively studied via Monte Carlo (MC) methods [4, 5, 6, 7]. Earlier studies reported estimates for α that are significantly lower compared to the LS value. Later, results obtained via certain extrapolation of the finite-time exponents were explained by arguing that the LS exponent can be realized only when $t \to \infty$ [8]. Naturally, this suggests that there exist corrections to the scaling law at finite time. More recently, however, it has been demonstrated [9, 10, 11], via appropriate finite-size scaling (FSS) analyses, that the LS growth is realized even at finite times, ruling out [6, 9, 10, 11] the presence of the above mentioned strong corrections. This motivated us to apply such FSS analyses to the study of kinetics in mixtures having more than two species, for which the number of studies is limited [12, 13, 14].

Basic physics of a multicomponent mixture can be captured by the q-state Potts model. For nearest-neighbor interaction the Hamiltonian for this model is (σ_i being the state at site i) [15]

$$H = -J \sum_{\langle ij \rangle} \delta_{\sigma_i,\sigma_j}; \;\; \sigma_i = 1, 2, \ldots, q, \tag{1}$$

where $\delta_{m,n}$ is the Kronecker delta function and $J > 0$ is the interaction strength. For $q = 2$, the Hamiltonian in Eq. (1) reduces to the Ising one, apart from a prefactor. This model undergoes a phase transition which is of second order for $q \leq 4$ and beyond that it is of first order [15]. The transition temperature of the model is given by $T_c(q) = J/\left[k_B \ln(1+\sqrt{q})\right]$, with k_B being the Boltzmann constant. In this article, we extend our previous work [16] on the segregation kinetics in multicomponent mixtures to higher values of q via MC simulations of the above model, subsequent analyses of the data being performed via a FSS method.

2. Methods

We perform canonical MC simulations [17] in space dimension $d = 2$, on square lattices, with periodic boundary conditions. Since the number of particles of each type, during phase separation, is constant, the simulation is conducted via the standard Kawasaki dynamics [17]. In this method, a trial move is performed by interchanging positions of two nearest-neighbor particles, chosen at random. The standard Metropolis criterion is used to accept such moves [17]. Our unit of time is one MC step (MCS) that consists of L^2 attempted moves where L is the linear dimension of the square lattice. An initial configuration, mimicking a high-temperature homogeneous mixture, consists of a random arrangement with equal concentration of all spin or particle states. Finally, the simulations are run with $L = 128$, at quench temperatures $T < T_c(q)$. In our simulations the unit of T is J/k_B, and we set both J and k_B to unity. All quantitative results are presented after averaging over 50 independent runs.

3. Results

Evolution of systems, following temperature quenches, are depicted in Fig. 1(a), where we show snapshots from two different times, for a small q (= 3) as well as a large q (= 20). For high q, especially at early times, there exists a large number of point defects which are defined as the meeting point of three or more states. However, at late times, even for q as high as 20, the one-dimensional line defects or domain walls are the majority, and therefore, the growth is not expected to be affected by those point defects.

Structural self-similarity, that is generally associated with coarsening phenomena, is reflected in the superposition properties of a number of morphology related functions such as the two-point equal-time correlation, the structure factor, and the size distribution function of the domains [1]. For each of these functions, data from different times t can be collapsed onto appropriate master curves that are independent of t [1]. For example, the distribution function, $P(\ell_d, t)$, of domain lengths ℓ_d, obeys the scaling $P(\ell_d, t) = \ell(t)^{-1}\tilde{P}[\ell_d/\ell(t)]$, with $\tilde{P}$ being a master function [1, 2]. In this work we rely on this scaling for consistent self-similarity during growth [16] and obtain the average domain size as $\ell(t) = \overline{\ell}_d = \int d\ell_d \, \ell_d P(\ell_d, t)$, where ℓ_d is measured from the distance between a pair of neighboring domain boundaries, along any Cartesian axis. To circumvent the effect of thermal noise in estimating ℓ_d, we apply a majority spin rule that provides pure, zero temperature like, domain structures [10]. The dependence of ℓ on t, for a few values of q, at $T = 0.7T_c(q)$, are shown in Fig. 1(b). The inconsistency [16] of the data with the expected LS growth (shown by the solid line) increases with q. Such discrepancy was previously [8, 14] attributed to the presence of finite-time corrections and it was believed that the LS growth can only be realized for $\ell(t) \to \infty$. In the following, via application of FSS, we show that those corrections are negligible and the LS law can be observed at finite times, even for very large q.

By incorporating an initial domain size ℓ_0, which is the average length at time t_0 (measured from the moment of quench) when the mixture renders unstable to fluctuations, we quantify the growth as [9, 10, 18, 11]

$$\ell(t) = \ell_0 + A\,(t - t_0)^{\alpha}. \tag{2}$$

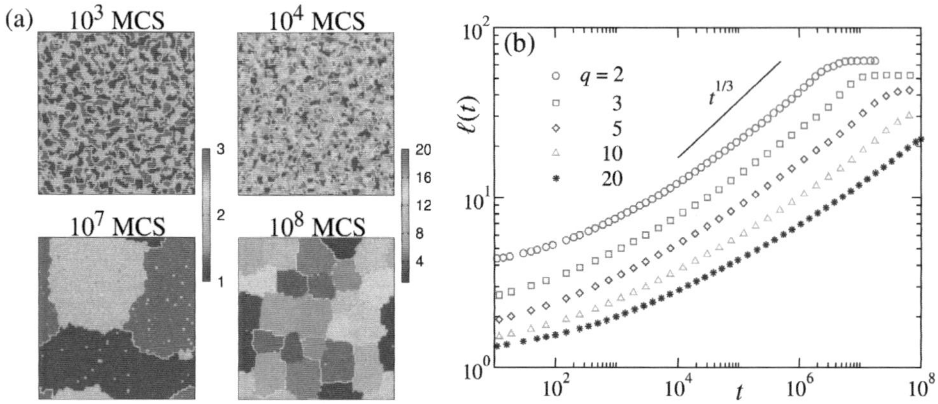

Figure 1. (a) Snapshots showing domain coarsening in the $d = 2$ Potts model with $q = 3$ (left) and $q = 20$ (right) at $T = 0.7T_c(q)$, following quenches from randomly mixed phases at high temperature. Color coding for different states are separately shown for each of the q values. (b) Log-log plots of the characteristic length $\ell(t)$ versus t, for five values of q, at $T = 0.7T_c(q)$. The solid line represents the LS growth.

Recalling FSS in equilibrium critical phenomena [19, 20], one uses the correspondence $\ell(t) - \ell_0 \hat{=} \xi$ and $1/(t - t_0) \hat{=} |T - T_c|$, to write [18]

$$Y(y) \;=\; \frac{\ell(t) - \ell_0}{\ell_{\max} - \ell_0}, \text{ choosing } y = \frac{(\ell_{\max} - \ell_0)^{1/\alpha}}{(t - t_0)}. \tag{3}$$

Note that $\ell_{\max}$ ($\sim L$) is the saturation length, i.e., the largest domain size one can access in a finite system of size L [9, 16]. From Eqs. (2) and (3), it transpires that in the growth regime ($y \to \infty$) $Y(y) \sim y^{-\alpha}$, and when one approaches the finite-size limit ($y \to 0$) $Y(y) \to 1$. During the FSS analysis α is tuned, with the objective of superimposing data from different L optimally. For the best collapse, Y should obey the above limiting forms. We obtained values of ℓ_0 (or t_0) from $\alpha_i = \frac{d \ln \ell(t)}{d \ln t}$, the finite-time exponents [8], versus $1/\ell(t)$ data (see Ref. [16] for details), and so, these are not adjustable parameters. For brevity, we do not present any result on FSS for a fixed q, instead move onto capturing the universality in the scaling for different q.

For that purpose we modify the scaling variable y by rewriting it as [16]

$$y_q = f_q \frac{(\ell_{\max} - \ell_0)^{1/\alpha}}{(t - t_0)}, \text{ where } f_q = \left[\frac{A(q = 2)}{A(q)}\right]^{1/\alpha}. \tag{4}$$

The metric factor f_q is related to the amplitude A in Eq. (2), which is a function of q. We have $Y = 1$, as $t \to \infty$, for all q. Thus, without f_q, for different values of q, the $Y(y)$ in Eq. (3) would only have horizontal separation from each other. Here, treating $q = 2$ as reference, to obtain an optimum collapse, along with the $Y(y_q) \sim y_q^{-\alpha}$ behavior, we adjust f_q for data from different q. Reasonably good collapse of data, that is obtained by fixing $\alpha = 1/3$ [shown in Fig. 2(a)], confirms not only the LS growth from very early time for all q, but also the fact that the growth dynamics, irrespective of q, is describable by a single universal FSS function. This combined fact further implies that finite-size effects are also the same for all q. This is consistent with the report of very robust universality in finite-size effects in coarsening phenomena, observation made from studies of other systems [11, 18]. Existence of such universal FSS function got reported first in connection with equilibrium critical phenomena [21, 22] and has recently been

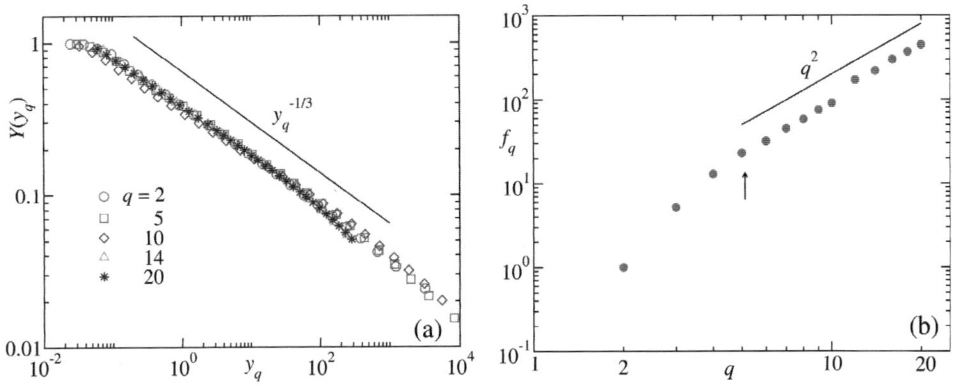

Figure 2. (a) Illustration of the universality of the FSS function, $Y(y_q)$, for the domain growth for different q at $T = 0.7T_c(q)$, extending Fig. 10 of Ref. [16] to larger values of q. Here $Y(y_q)$ is shown versus the scaling variable y_q defined in (4) that contains a nonuniversal metric factor f_q. The solid line corresponds to the expected $Y(y_q) \sim y_q^{-\alpha}$ behavior for $y_q \gg 0$. (b) Log-log plot of f_q as a function of q. The solid line represents q^2 behavior. The arrow points towards $q = 5$.

adapted to understand the kinetics of polymer collapse [23, 24]. As evident, even though the scaling function is universal, f_q is nonuniversal. Its variation as a function of q, plotted on a log-log scale in Fig. 2(b), shows a crossover at $q = 5$, the onset point of first-order transitions in the $d = 2$ Potts model. The post-crossover behavior of f_q is quite consistent with q^2, shown by the solid line. More comprehensive studies are essential to validate this observation [25].

For the temperature dependence of the growth too, one can design a similar scaling apparatus, by replacing y_q with [16]

$$y_T = f_T \frac{(\ell_{\max} - \ell_0)^{1/\alpha}}{(t - t_0)}, \text{ where } f_T = \left[\frac{A(T = 0.7\,T_c)}{A(T)}\right]^{1/\alpha}. \tag{5}$$

Again, one needs to look for the optimum collapse of data, in this case from different T, for a fixed q, along with the corresponding $Y(y_T) \sim y_T^{-\alpha}$ behavior in the early time regime. Here also fixing α to $1/3$ yields reasonable quality of collapse of data for $T \in [0.5T_c(q), 0.85T_c(q)]$, confirming the temperature independence of the LS growth. For details of this exercise and corresponding scaling plots see Ref. [16].

Finally, combining Eqs. (4) and (5) one can introduce a "super"-scaling variable [16]

$$y_s = f_q f_T \frac{(\ell_{\max} - \ell_0)^{1/\alpha}}{(t - t_0)}. \tag{6}$$

Using y_s it can be easily shown that the FSS function is not only universal but rather superuniversal. This fact is nicely demonstrated in Fig. 3 where the scaling plots coming from data for different q and T are made to collapse onto a single master curve, along with the realization of $Y(y_s) \sim y_s^{-\alpha}$ behavior.

4. Conclusion

In this paper, we have extended our finding [16] of a universal FSS function for the kinetics of phase separation in the two-dimensional q-state Potts model to higher values of q (up to $q = 20$). Although there exist many more point defects for higher q, the observation of universality of the FSS function, for all q, along with the Lifshitz-Slyozov behavior, suggests that the growth is

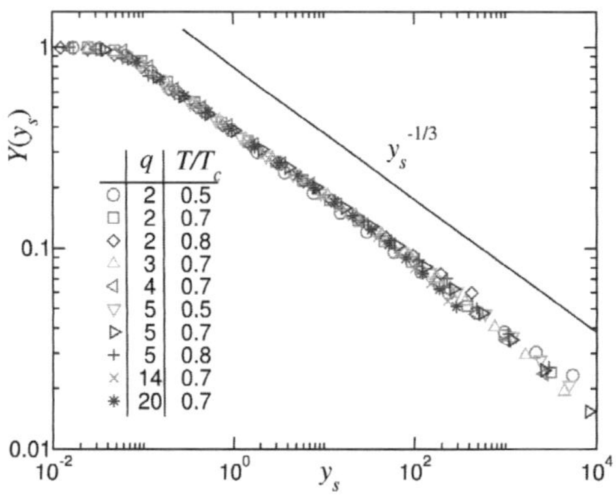

Figure 3. Plot elucidating the fact that domain coarsening in the q-state Potts model is describable by a superuniversal finite-size scaling function $Y(y_s)$, independent of q and the quench temperature T. The solid line has the power-law form $Y(y_s) \sim y_s^{-\alpha}$.

indeed driven by domain wall shrinking. Interestingly, we have observed signature of a crossover in the behavior of the metric factor f_q at $q = 5$ that coincides with the onset point for first-order transitions in the Potts model. A thorough investigation of this connection is needed. It remains to be seen if an analogous study with nonconserved dynamics provides similar information [25].

Acknowledgments
The work was funded by the Deutsche Forschungsgemeinschaft (DFG) under Grant No. JA 483/33-1. We obtained further support from the Deutsch-Französische Hochschule (DFH-UFA) through the Doctoral College "$\mathbb{L}^4$" under Grant No. CDFA-02-07 and the European Union through the FP7 Marie Curie IRSES network DIONICOS under Contract No. PIRSES-GA-2013-612707.

References
[1] Bray A J 2002 *Adv. Phys.* **51** 481
[2] Puri S and Wadhawan V eds 2009 *Kinetics of Phase Transitions* (CRC Press, Boca Raton)
[3] Lifshitz I M and Slyozov V V 1961 *J. Phys. Chem. Solids* **19** 35
[4] Rao M, Karlo M H, Lebowitz J L and Marro J 1976 *Phys. Rev. B* **13** 4328
[5] Roland C and Grant M 1988 *Phys. Rev. Lett.* **60** 2657
[6] Amar J G, Sullivan F E and Mountain R D 1988 *Phys. Rev. B* **37** 196
[7] Marko J F and Barkema G T 1995 *Phys. Rev. E* **52** 2522
[8] Huse D A 1986 *Phys. Rev. B* **34** 7845
[9] Majumder S and Das S K 2010 *Phys. Rev. E* **81** 050102(R)
[10] Majumder S and Das S K 2011 *Phys. Rev. E* **84** 021110
[11] Majumder S and Das S K 2013 *Phys. Chem. Chem. Phys.* **15** 13209
[12] Grest G S and Sahni S P 1984 *Phys. Rev. B* **30** 226
[13] Jeppesen C and Mouritsen O G 1993 *Phys. Rev. B* **47** 14724
[14] Das S K and Puri S 2002 *Phys. Rev. E* **65** 026141
[15] Wu F Y 1982 *Rev. Mod. Phys.* **54** 235
[16] Majumder S, Das S K and Janke W 2018 *Phys. Rev. E* **98** 042142
[17] Landau D P and Binder K 2009 *A Guide to Monte Carlo Simulations in Statistical Physics* (Cambridge University Press, Cambridge)
[18] Das S K, Roy S, Majumder S and Ahmad S 2012 *Europhys. Lett.* **97** 66006
[19] Fisher M E 1971 *The theory of critical point singularities* in *Critical Phenomena* ed Green M S (Academic Press, London), p. 1
[20] Privman V ed 1990 *Finite Size Scaling and the Numerical Simulations of Statistical Systems* (World Scientific, Singapore)
[21] Privman V and Fisher M E 1984 *Phys. Rev. B* **30** 322 (1984)
[22] Hu C-K, Lin C-Y and Chen J-A 1995 *Phys. Rev. Lett.* **75** 193
[23] Majumder S, Zierenberg J and Janke W 2017 *Soft Matter* **13** 1276
[24] Christiansen H, Majumder S and Janke W 2017 *J. Chem. Phys.* **147** 094902
[25] Majumder S, Das S K and Janke W work in progress

Three dimensional visualization of simulations of liquids and solids

Joan Adler, Eden Segal and Grisha Zeltyn

Physics Department, Technion-IIT, Haifa, Israel, 32000

E-mail: phr76ja@technion.ac.il

Abstract. Visualization in three dimensions is invaluable for understanding the nature of condensed and fluid systems, but it is not always easy. In nature it is hard to view sample interiors, but on computers it is possible. We describe and contrast two opposite approaches - "smoke" visualization for viewing interiors of liquid samples and interactive WebGL for solids and molecules. Both are extensions of earlier Technion Computational Physics group projects and complement and are interoperable with the recent SimPhoNy Fp7 project. They require only desktop hardware and software accessible to students. Examples and standalone instructions for both are presented, starting with sample creation and concluding with image galleries.

1. Introduction

Let us ask "what is the best way to look into our three-dimensional world? Virtual reality is cool; for movies, gaming and research/teaching. Fortunately the first two spur software and hardware development that we can use for research and teaching. But they also spoil us with heavy expectations. There are many visualization codes for research/teaching - for a recent discussion of several that can interface well with simulation codes see [1]. The SimPhoNy project provided a way to connect simulations at different scales with visualization software, and provides an approach to integrating simulation codes with visualization ones in a standardised, repeatable framework. Its novelty was in the interoperability, rather than in new visualization approaches. Good documentation was also a focal point and wrappers to pass data between different codes were presented. We demonstrated interoperability [1] by showing the same *.xyz* data for electronic density with three diferent codes. This demonstration showed that concepts of color use, persepective and so on are the crucial aspects, and code selection secondary.

We now present two further developments in a similar spirit based on the *.xyz* datafiles of the AViz code [2, 3, 4] One is an approach extending the AViz code to density simulation such as that generated from 3D Lattice Boltzman (LB) fluid simulations [5], and the other an implementation of WebGL [6] for nanotubes and crystals, extending earlier AViz studies. Both can be carried out on commodity boxes. Supplementary material relating to these projects is presented on [7].

Very large LB simulations [8] were part of SimPhoNy, but their visualization was carried with Computational Fluid Dynamics (CFD) codes and an HPC center visualization expert. We now describe a much simpler 3D LB simulation and its visualization with AViz which enables insight into the flow process from every desktop [9]. In the next section we will discuss our

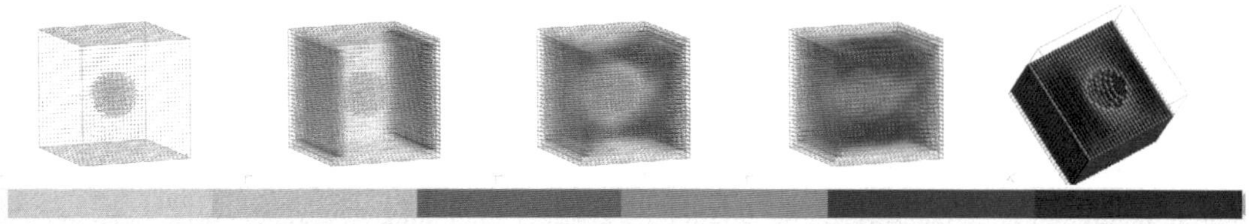

Figure 1. A series of flow stages with spin objects: empty, start of flow, 11th timestep, 29th timestep, with the color bar that ranges from 0.0 to 6.0 nodes/iteration step. An image of a sliced sample with ball objects near the 29th timestep is at the extreme right.

previous approaches and then introduce a specific LB implementation that enables a novel approach to the 3D visualization of the flow. We will implement in AViz, but again the data files could be viewed with other codes. Following this, we will describe a new WebGL approach to nanotube modelling, again demonstrating interoperability with earlier AViz visualizations. Final discussion will center on the question "is it worth the effort?"

2. LB simulations and their websites

LB is an innovative approach to solving the Boltzmann equation, a partial differential equation governing the mass distribution of different fluid particles as a parameter of time, position and velocity. The equation is based on the assumption that different fluid particles are independent of each other except for collisions between them. The LB simulation quantizes each of the parameters: time is incremented in discrete steps and the positions can only fall on a grid of points. It is an example of a cellular automaton. A website simulation/visualization approach to cellular automata was presented in [10].

The novelty in this project is an educational approach with reproducible code for a commodity box. There are more sophisticated calculations in the literature [8]. Our local efforts have been focused on instructional websites and public codes [11, 12], details in the supplementary material [7] In [12], a GPU was also utilised in the visualization leading to an elegant, fast code that was somewhat demanding in terms of hardware and software requirements. The project code is not split into simulation and visualization parts; and the specific system demands (GPU and OpenGL, GLUT and GLEW) limit its future implementation, although the code is perfectly reproducible on suitable hardware and has been checked on different LINUX implementations. It includes a text-file method to create the porous structure in any desired shape. That hint of turbulence led to the idea of improved visualization techniques in order to observe this more closely, and in three dimensions.

3. Three dimensional density visualization with AViz

There are several approaches to 3d density visualization. Assuming that we do not have a convenient immersive cave; one good approach used for LB [8] is to show contour plots at selected locations and leave it to the viewer to interpolate between these. Another is to use "smoke" visualization with and without a stereo option. A third is to combine "smoke" with strategic slicing. We have explored the latter two choices for LB.

The separation of simulation and visualization enables a choice of visualization code. As noted above, in [1] we created visualizations for the same electronic density data in three different codes with the crucial issue being approach (as in selecting which parts of the data to show, rather than in the code selection). We direct the interested reader to electronic density visualizations in Figures 8,9, and 10 of [1, 7].

4. Three dimensions for LB

The present LB simulation [5] overcomes the size and non-public domain limitations of the first [11], and the independent data access issues of the second [12] as well as the dimensional limitation of both but does not have a general porous shape selection option - only several preselected geometrical shapes. It is three dimensional; and although it works best on specific hardware (SIMD, details below) it can also be run on simpler systems and even has a Windows capability.

These present calculations follow the paradigm of splitting the effort into simulation and visualization, as per our group preferences and we present details of both. Downloadable code and all details can be found on the LB project website [5]. Like other LB simulations there are several stages in the simulation algorithm. The first is collision, including calculation of velocity and density, stored for later use in visualization. Then "streaming" or propogation to the next cell. Less straightforward is the fluid wall interaction, where particles reverse when they hit a wall. Boundary conditions, both at walls and inlet/outlet is the hardest part, the present code uses Zho-He conditions with details on the website.

The 3 dimensional algorithm's choice of 3 dimensions and 19 velocity values is given the name Lattice Boltzman D3Q19 or D3Q19 for short. The velocities for this case take one of 19 different values, which together with implementing inlet and outlet boundary conditions even for $\tau \neq 1$ leads to a somewhat complicated implementation, but it is documented in detail on the website. τ is an inverse weight.

Hardware and software details and downloadable code are given on the website. The simulation has been designed to run on two levels of hardware specifications; the simpler non-SIMD one for all processors requires either gcc4.8 or gcc4.9 and the one suited for SIMD processors needs gcc4.9 Both use cpp support. SIMD (Single Instruction for Multiple Data) is present on Intel architectures newer than Haswell (fron 2012) and AMD newer than Piledriver (from 2013). Multithreading with the OpenMP library that is part of the compiler makes the simulation even faster, because it distributes both the streaming and collision parts to the different threads in the processor. In the collision step, the distribution is carried out based on the z axis for the collision step, each thread is given some number of identical z planes. On the streaming step the distribution is based on the direction.

Another factor that increases the thruput is the memory layout which is an SOA (structure of arrays) approach. For each cell in our simulation there are 19 different values needing to be stored, one for each direction. We chose to store the values in 19 different arrays. This approach was used for the 2 simulation buffer, one for post streaming and one for post collision, and for the velocity buffers (which use only 3 arrays).

This SOA approach is very helpful in the SIMD case. SIMD means that each instruction works on a vector register of 256 bits, which are 8 floats, 4 doubles or some number of integers based on the size. Each instruction is performed on each float of the array independently on any other vectors element. For those instruction to be effective, the programmer needs to make each vector element be as similiar to the other elements as much as possible. Because each of the pixels in the collision performs the same instruction irrespectively of the other pixels, making each vector element be a different pixel is a perfect candidate for this instructions. Loading such registers is best done by loading adjacent values in memory, and that is the reason we use the SOA order. The collision step operates on 8 pixels at the same time, each of those share the same y and z coordinates but differ in their x, and are then stored in the same way. The streaming step is done in exactly the same way. Detailed descriptions, sourcecode and pseudocode are given on the website.

The splitting of the simulation and visualization steps means that a fast processor can be used relatively easily for the heavier simulation part, and the hardware/software demanding visualization part be implemented on older but well-installed desktops at leisure and repeated

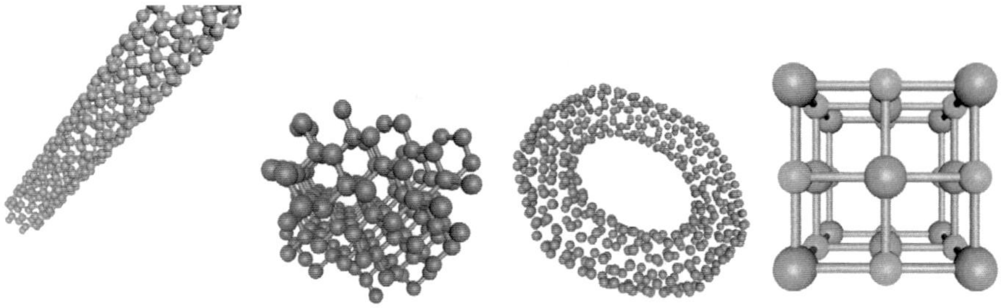

Figure 2. Nanotube, diamond, C_{440}, NaCl

as needed. The code generates data files in the *.xyz* format every 10 steps as a default.

As described above, for visualizations decisions on what to show, emphasize, manipulate, rather than which code is used are the most important so w.l.o.g. we use AViz. We aim to view "into" a semitransparent sample implementing both color binning [13] and slicing [14] where appropriate. We also use techniques to aid three dimensional insight including viewing an initial empty sample to establish perspective, and animations. An example of the empty box is given at left in the Figure 1, and three later shots center and right. AViz has several object options. For the entire sample shot we used the spin option for the fluid that shows orientation and direction with objects small enough to enable transparency. For the slicing we selected a "ball" object option so that the plane of the cut could be clearly viewed when sliced, although opaque when closed. The static cells are smaller balls. In both situations the same color bar is used, with static elements in blue or absent.

For these runs the sample dimension is 64*60*60 nodes, with a ball of radius 15 in the middle, and walls in the four enclosing planes perpendicular to the y and z axis. The boundary condition is that both the inlet speed and the outlet speed is 0.02 nodes/iteration time with τ=1.0.

5. WebGL for nanotubes and crystals

This is an extension of a WebGL site [16] (http://phony1.technion.ac.il/~lsharir) that provides square and cubic lattice percolation visualizations for nanotubes and other structures with uniform bond lengths. The nanotube generation is based on an AViz nanotube website [17] and includes a generalization to all lattices with nearest neighbour bonds. WebGL is a version of the graphic library OpenGL viewed thru browsers that avoids the complications of Open GL installation and can be viewed with browsers on smartphones and tablets. It is based on javascript, rather than the c/c++ basis of OpenGL and AViz. In common with AViz (that a uses OpenGL) our WebGL sites are also require/create .xyz files, and the limitations on WebGL (relative to the OpenGL/mesa implementation used for AViz) do not affect the visualizations presented below.

The AViz nanotube creation site is based on the [18] nanotube polyhedral method that correctly describes nanotube curvature. The code on the site enables creation of three types (Carbon, Boron and Silicon) of .xyz files for nanotubes with any desired parameters. These files have to be transformed for the WebGL input using the ChemDoodle Web Components Library [20] and a MATLAB script is provided for this. A translation to c/c++ is in preparation. Some still images [6] are presented in Figure 2 but even more so than for AViz images the beauty is in the rotate and zoom possibilities, which require a browser. They can be viewed on the website [7].

6. Conclusions

These, and other visualization techniques purport to enable relatively effortless three-dimensional visualization into sample interiors, the former for fluids, the latter for molecules and crystals. Considerable effort was made to develop both websites with clear instructions for users. For the former LB + AViz code, (AViz has validated instructions for systems programmer installation, [2, 3]) and a robust makefile for the LB code on a LINUX desktop are the sole requirements. The WebGL site is self contained but for now requires MATLAB, an internet connection to the ChemDoodle site and a source of *.xyz* files. The images can be shown in any browser, including that on smartphones. The *.xyz* files are a bit more elusive. Nanotubes [17] can be self-generated on demand, a set of links for other materials [22] and a request page [23] are provided. There are also many *.xyz* files searchable on the web.

Given the potential benefit from these types of visualizations and the simplicity of the desktop orientation, they are well-worth the additional small effort. A selection of animated movies are presented in [5] and in our supplementary material, mirrored at [7].

Acknowledgments

The educational websites were part of the SimPhoNy project which was funded by FP7 under NMP-2013-1.4-1 call with grant agreement no: 604005. We thank Keijo Mattila, Liran Sharir and Simon Brandon for discussions and advice.

References

[1] Adler J *et al.* 2018 Visualization in the integrated SimPhoNy multiscale simulation framework *Computer Physics Communications* **231** 45-61
[2] http://phony1.technion.ac.il/~aviz
[3] https://github.com/simphony/AViz
[4] Adler J 2003 Visualization in atomistic and spin simulations, *Computers in Science and Engineering* **5** 61-65
[5] http://phelafel.technion.ac.il/~edensegal
[6] http://phelafel.technion.ac.il/~gzeltyn
[7] http://phony1.technion.ac.il/~phr76ja/vin3iop/mirror.html
[8] Mattila K, Puurtinen T, Hyväluoma J, Surmas R, Myllys M, Turpeinen T, Robertsén F, Westerholm J and Timonen J 2016 A prospect for computing in porous materials research: Very large fluid flow simulations *Journal of Computational Science* **12** 62-76
[9] Peled D, Silverman A and Adler J 2013 3D visualization for atomistic simulations on every desktop **IOP Conference Series 454** 012076
[10] http://phelafel.technion.ac.il/~idokesten/pro/about/about.htm
[11] http://phelafel.technion.ac.il/~drorden/project
[12] Adler J, Nissim G and Kiswani A 2017 GPUs in a computational physics course", *J. Phys.: Conf. Ser.* **905** 012017
[13] Adler J, Fox J, Kalish R, Mutat T, Sorkin A and Warszawski E 2007 The essential role of visualization for modeling nanotubes and nanodiamond, *Computer Physics Communications* **177** 19-20
[14] Adler J and Pine P 2009 Visualization techniques for modelling carbon allotropes *Computer Physics Communications* **180** 580-2
[15] Grosso B, Cooper V R, Pine P, Hashibon A, Yaish Y and Adler J, 2015 Visualization of electronic density *Computer Physics Communications* **195**, 1-13
[16] Adler J, Elfenbaum S and Sharir L 2017 *Eur. Phys. J. Special Topics* **226** 737–747
[17] Mazvovsky D, Halioua G and Adler J 2012 *Physics Procedia* **34** 1-5
[18] Lee R K F, Cox B J and Hill J M 2010 *Nanoscale* **2** 859-872
[19] Adler J, Artzi Y, ben Bashat L, Izraeli T Y, Kreif M, Lavi I, Leibenzon A, Levi A, Schlesinger I, Toledano E, Peretz U, Weisler Y and Alon Yagil 2014 *Journal of Physics: Conference Series* **510** 012003
[20] http://web.chemdoodle.com
[21] http://phony1.technion.ac.il/~aviz/xyzfiles.html
[22] http://phony1.technion.ac.il/~phr76ja/plugfest/
[23] http://phony1.technion.ac.il/~aviz/requests.html

Journal of Physics: Conference Series **2122** (2021) 012011 doi:10.1088/1742-6596/2122/1/012011

Modeling electron transfer from the barrier in InAs/GaAs quantum dot-well structure

I Filikhin[1], Yu B Kuzmichev[2], V Mitic[3] and Th. Peterson[1], B Vlahovic[1]

[1] North Carolina Central University, CREST Center, 1801 Fayetteville St., 27707 Durham, USA

[2] Yaroslavl State Pedagogical University, 108 Respublikanskaya St., 150000, Yaroslavl, Russia

[3] University of Niš, Faculty of Electronic Engineering, Aleksandra Medvedeva 14, Niš, Serbia

E-mail: ifilikhin@nccu.edu

Abstract. We study single electron tunnelling from the barrier in the binary InAs/GaAs quantum structure including quantum well (QW) and quantum dot (QD). The tunneling is described in the terms of localized/delocalized states and their spectral distribution. The modeling is performed by using the phenomenological effective potential approach for InAs/GaAs heterostructures. The results for the two and three-dimensional models are presented. We focus on the effect of QD-QW geometry variations. The relation to the PL experiments is shown.

1. Introduction

Weakly coupled binary semiconductor nanoscale systems demonstrate perspectives for nano-sensor applications due to high sensitivity of electron localization and resonance tunnelling between the objects of the system on symmetry violation [1]. The single electron tunneling properties of one dimesional (1D), 2D and 3D structures as well as double quantum wells (DQWs), double quantum dots (DQDs), and quantum rings (DQR) are well known. The electron spectra of such quantum objects in two- and three dimensional (2D and 3D) were studied in previous works [2] with relation to the electron localizations and tunneling between the objects. The wave function of electron may be localized in one of the QDs or be delocalized when it is spread over the whole system. Tunneling occurs in the last case. In the condition of weak coupling objects, electron wave function can be localized in the both objects but with different probability. We explore experimental possibility for optical registration of electron localizations in binary quantum systems. In this work we focused on the resonance tunneling in 3D/1D and 2D/1D nanoscale InAs/GaAs dot-well complex. This complex has the mixed spectral structure: discrete spectrum for QD (in 3D and 2D) and continuous spectrum for QW when this QW is considering in three- or two-dimensional space [3, 4, 5]. Our modeling carrier transfer from the barrier in InAs/GaAs dot-well tunnel-injection structure is performed by using the band gap model based on the effective potential [6, 7]. It has to be stressed that the tunneling is described in the terms of localized/delocalized states and their spectral distribution [2]. We have simulated the main effect of the optical PL experiments given by Yu. I. Mazur et al. in [8].

2. Model

Presented work was motivated by the experimental results of coherent coupling between photo-excited quantum-dots and a quantum well reported in [8, 9, 10, 11, 12]. The schematic explanation the PL experiments [8] is given in Fig. 1 and Fig. 2. The coupling in the QD-QW complex was investigated in [8] by the dependence of quantum-dot photoluminescence as a function of quantum dot-well barrier thickness. It was shown that the resonant tunnelling rate is varied by changing the effective barrier thickness. This strongly affects the exciton dynamics in these hybrid structures as compared to isolated QW or QD. A specific behaviour of the communication between a quantum-dot and quantum wall was discussed in [11, 12].

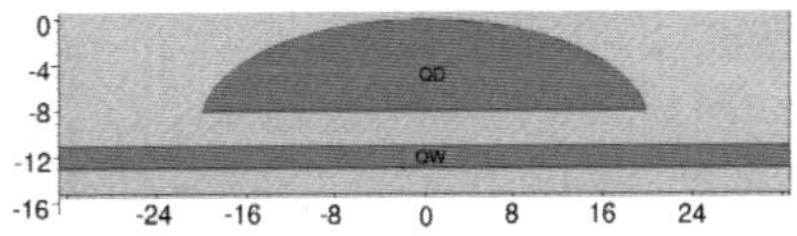

Figure 1. (Left) Geometry of the QD-QW complex (2D model). The sizes are given in nm. The profile of the fabricated QD is presented in [10].

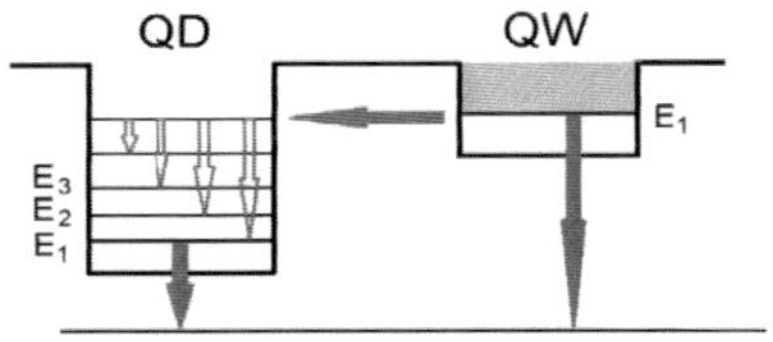

Figure 2. Schematic description of the PL experiments from [8]. Electron confinement structure and energy levels. The tunneling between QD and QW is shown by horizontal arrow.

The InAs/GaAs quantum heterostruction is modelled using the effective potential model. The effective potential simulates the strain effect in the InAs/GaAs heterostructures [6, 7]. The band gap model for description of the QDs (and QWs) is presented in Fig. 3. Here, the effective potential is given by V_s and the bang gap potential is V_c.

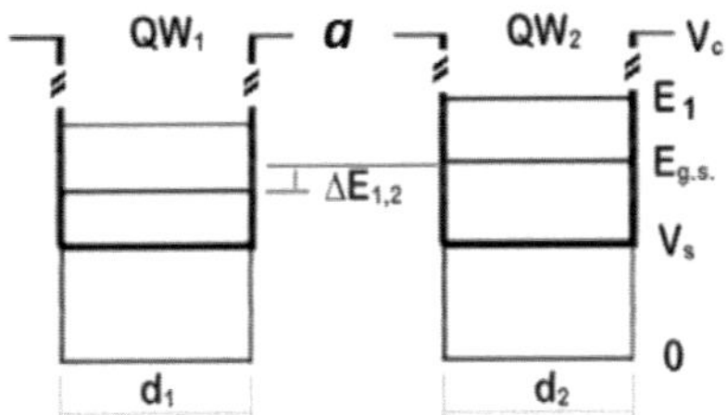

Figure 3. Band gap model for InAs/GaAs double quantum well (in 1D). The system includes two quantum wells, QW_1 and QW_2, which are differ by the sizes d_1 and d_2. Inter-well distance is a.

We consider the QD-QW complex as two-level quantum system [13] and describe the electron tunneling in the system within terminology of spectral distributions of the localized/delocalised states [1].

Double quantum dot is an example of the two-level quantum systems. The spectrum of single electron confinement states in DQDs is a set of quasi-doublet levels. The electron can be localized in one of the quantum dots or be delocalized over both the QDs. The electron tunnelling between QDs in DQD is indicated by changing of the type of electron localization. The tunneling occurs through the anti-crossing of levels and, in case identical QDs, extremely sensitive on shape symmetry violations in the systems [1, 2]. To evaluate the electron localization, one can use the electron average coordinate $\langle x \rangle$ which is calculated as: $\langle x \rangle = \langle \Psi | x | \Psi \rangle$, where Ψ is wave function of the system. In the approach of the two-level system, Ψ is represented by superposition of the wave functions ψ_1 and ψ_2 of separated left (1) and right (2) quantum dots. The x-coordinate origin is the mid-point of the two QWs. The average coordinates $\langle x \rangle_+$ and $\langle x \rangle_-$, for the electron wave functions of the quasi-doublet, Ψ_+ and Ψ_-, can be written as

$$\langle x\rangle_+ = -\cos^2\left(\frac{\Theta}{2}\right)|\langle x\rangle_1| + \sin^2\left(\frac{\Theta}{2}\right)|\langle x\rangle_2| + 2\sin\left(\frac{\Theta}{2}\right)\cos\left(\frac{\Theta}{2}\right)\langle x\rangle_{12},$$

$$\langle x\rangle_- = -\sin^2(\Theta/2)\,|\langle x\rangle_1| + \cos^2(\Theta/2)\,|\langle x\rangle_2| - 2\sin(\Theta/2)\cos(\Theta/2)\,\langle x\rangle_{12},$$

where, $\Theta = arctg\,(W/\Delta E_{12})$. The parameter W is a coupling coefficient of the quantum system elements. It depends on the wave function overlap for the "unperturbed states" ψ_1 and ψ_2. $\Delta E_{12} = E_1 - E_2$ is energy difference of separated QWs as is shown in Fig. 3. The matrix element W (which is also proportional to the quasi-doublet energy splitting ΔE) can be described using the following relation [14]: $W \sim S$, where S is the overlap integral, approximated by $S = \int_\Sigma \psi^2(x,y)\,dxdy$, with the integration domain Σ being the area between the QWs. It is clear, that the electron localization in DQW is extremely sensitive to small violations of DQW symmetry, when $\Delta E_{12} \approx 0$ [1].

3. Numerical modeling for carrier transfer from the barrier in InAs/GaAs–InGaAs/GaAs dot-well

The tunneling in the hybrid QD-QW complex is considered with relation to the PL experiments [8] (see Fig. 2). The QD - QW tunneling has been experimentally indicated by the existing PL peaks between lowest QD and QW peaks which were observed for separated QD and QW.

The 3D modeling is including the QD and QW having the rotation symmetry as is shown in Fig. 4(Right). Neumann boundary conditions were used for the QW. The results of numerical modeling are given in Fig. 4(Left) for the spectral distributions of the localized/delocalized states shown for several chooses of QD-QW geometry. The tunneling between QW and QD is occurred for the first energy level of QW. This situation directly comparable with the experimental picture proposed in Ref. [8]. The tunneling is possible for the close depositing of QD and QW. The

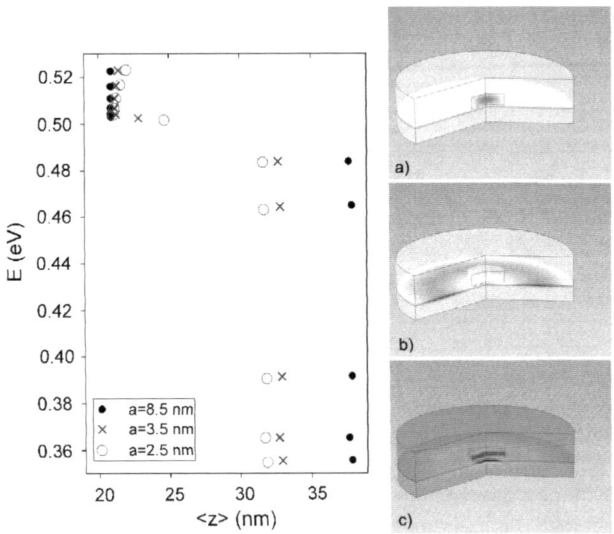

Figure 4. (Left) The spectral distribution of localized/delocalized states in 3D InAs/GaAs dot-well complex for different barrier thicknesses, a, shown by various symbols along the averaged coordinate $\langle z\rangle$. Electron tunneling occurs on the QW lowest energy level and relates to the delocalized state (noted by the arrow). (Right) The square of the wave function. The electron localized a) in QD, b) in QW and delocalized c) over the QD-QW complex.

dependence presented in Fig. 4(Left) shows that the barrier thickness a=8 nm is critical for the tunneling which is occurred when $a \ll 8$ nm and detected for $\langle z\rangle$ in diapason of [22,30] nm. When a>8 nm, the tunneling is not possible and we see the spectra of separated QD and QW. The spectrum of QD is discrete and electron localizes about $\langle z\rangle$ ≈38 nm. The QW spectrum is quasi-continuous due to numerical approximation and the electron is localized in QW with

$\langle z \rangle \approx$21 nm. In Fig. 4(Right) we illustrate the tunneling and show the electron wave functions for the different energy levels.

In Fig. 5, we present the results of numerical modeling for tunneling taken a place in the 2D-1D QD-QW complex. One can see in Fig. 5 (Left) that the QD and QW may be described as separated for large barrier thickness. The electron spectrum relates to single QD or QW. Corresponding localization is noted as "QD" or "QW" in the figure. The tunneling occurs for the lowest QW level (or higher confinement state of QD) when a is decreased to 8 nm and electron is localized in QD and QW, with $\langle y \rangle \approx$9 nm. This property is again directly related to the experiments [8]. Experiments given in [9] can be modeled by increasing the size of QD-QW

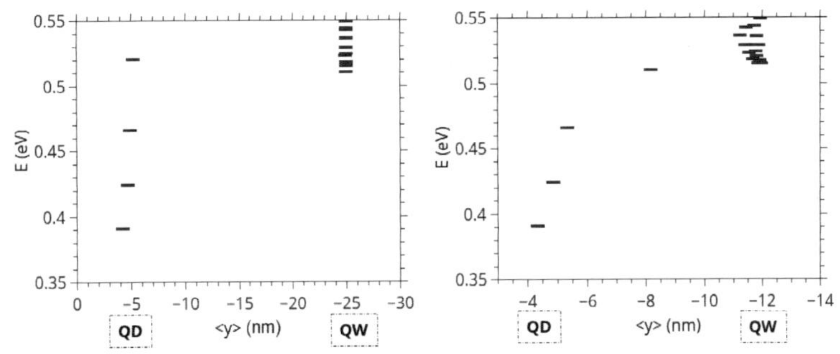

Figure 5. The spectral distribution of localized/delocalized states in 2D InAs/GaAs dot-well complex for different barrier thicknesses, a. (Left) a=20 nm and there are no delocalized (tunneling) states. (Right) a=8 nm.

complex. The schematic description of the PL experiments [9] is presented in Fig. 6.

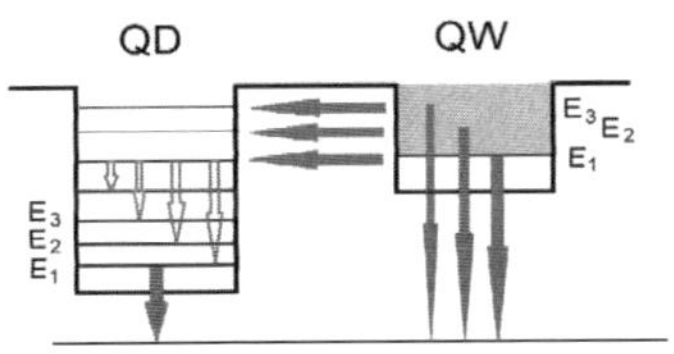

Figure 6. Schematic description of the PL experiments from [9]. Electron confinement structure and energy levels. The tunneling between QD and QW is shown by horizontal arrows.

The results of numerical modeling for tunneling taken a place in reconfigured 2D-1D QD-QW complex are presented in In Fig. 7. The calculations were performed for different value of the barrier thickness a. Once more, one can see that the tunneling is possible when the QD and QW are enough close. The tunneling is occurred through higher levels of discrete spectrum of QD and quasi-continuous spectrum of QW. Comparing this result with the QD-QW experiment one can conclude that the QW in experiment has quasi-continuous spectrum too. Our interpretation for this fact is that region of interaction between QD and QW is local and has a limitation by size of QD-QW interface B. The spectral level energy can be evaluated as L^2/B^2, where $L = 1, 2, 3, \ldots$ for each bands generated by the QW thickness.

4. Conclusion

We modeled the resonance tunneling in InAs/GaAs QD-QW complex including 2D (3D) and 1D quantum objects in terms of localized/delocalizad states of electron confinement spectrum according the formalism of two-level quantum systems. This simulations covers the main phenomenons of the optical PL experiments given by Yu. I. Mazur et al. in [8] and [9]. It was shown that the resonance tunneling spectral distribution depends on geometry factors of the QD-QW complex which can be predicted in theoretical simulations. The optical registration of electron localization in binary quantum systems can be used in new nono-sized devices.

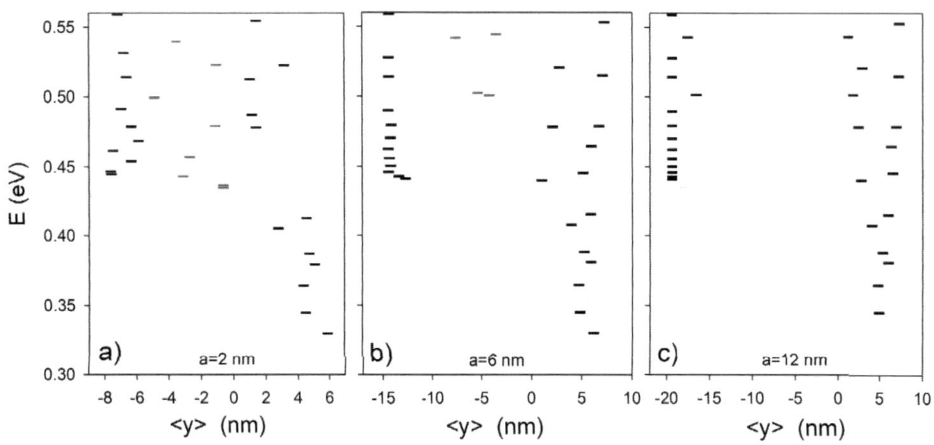

Figure 7. The spectral distribution of localized/delocalized states in 2D InAs/GaAs dot-well complex for different barrier thicknesses a) a=2 nm, b) a=6 nm and c) a=12 nm. Delocalized states are shown in red.

Acknowledgments
This work is supported by the RISE 1829245, DNS 2016-ST-062-000004 and DMR-1523617 awards.

References
[1] Filikhin I, A. Karoui A and Vlahovic B 2016 *Journal of Nanotechnology*, Article ID 3794109; doi:10.1155/2016/3794109
[2] Filikhin I, Matinyan S G and Vlahovic B 2015 *Quantum Matter* **4** 1
[3] Chuang S L and Holonyak N Jr 2002 Appl. Phys. Lett. **80** 1270
[4] L.I. Goray, P.N. Racec, *Proceedings of the International Conference DAYS on DIFFRACTION 2014*, pp. 89-95; IEEE: CFP14489-ART
[5] W. Rudno-Rudzinski et al. 2009 *Appl. Phys. Lett.* **94** 171906; DOI: 10.1063/1.3122935
[6] Filikhin I, Suslov V M, Wu M and Vlahovic B 2009 *Physica E* **41** 1358
[7] Filikhin I, Suslov V M and B. Vlahovic B 2006 *Phys. Rev. B* **73** 205332
[8] Mazur Yu I, et al. 2010 *J. Appl. Phys.* **108** 074316
[9] Mazur Yu I, et al. 2011 *J. Appl. Phys.* **98** 083118
[10] Mazur Yu I, et al. 2012 Nanoscale **4** 7509
[11] Mazur Yu I, et al. 2013 *J. Appl. Phys.* **113** 034309
[12] Guzun D, et al. 2013 *J. Appl. Phys.* **113** 154304
[13] Cohen-Tannoudji C, Diu B, Laloe F 1977 *Quantum Mechanics*, Volume 1, Wiley VCH
[14] Filikhin I, Matinyan S G and Vlahovic B 2014 Sensors & Transducers **183** 116

AUTHOR INDEX